人工智能与大数据系列

PyTorch
深度学习入门与技术实践

罗刚 / 著

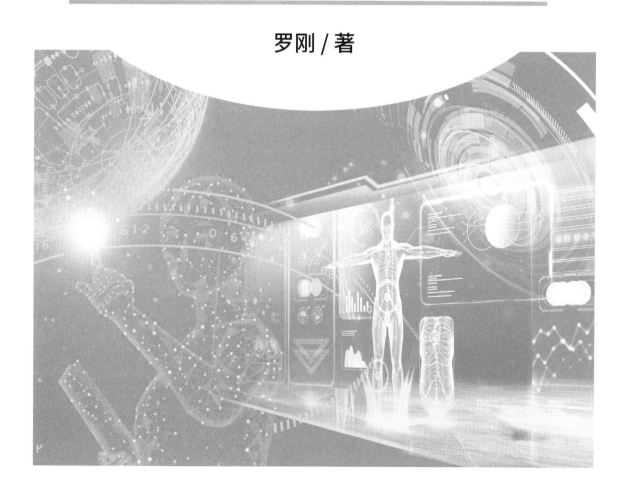

清华大学出版社
北 京

内容简介

本书介绍如何学习和使用流行的 PyTorch 框架开发深度学习应用，主要内容包括 PyTorch 中的计算图，用三阶多项式拟合函数，实现手写数字识别，神经网络基础，卷积神经网络，PyTorch 基础知识，transformer 架构，文本分类应用开发，聊天机器人应用开发，用 Wav2Vec2 进行语音识别，机器翻译应用开发，分布式 PyTorch 等。

本书适合作为高等院校计算机、软件工程专业本科生、研究生的参考书目，同时也适用于对 PyTorch 深度学习领域感兴趣的人士。

版权所有，侵权必究。举报：010-62782989，beiqinquan@tup.tsinghua.edu.cn。

图书在版编目（CIP）数据

PyTorch深度学习入门与技术实践 / 罗刚著.
北京：清华大学出版社，2025.2. --（人工智能与大数据系列）. -- ISBN 978-7-302-68180-9

Ⅰ．TP181

中国国家版本馆CIP数据核字第2025GJ9745号

责任编辑：张　敏
封面设计：郭二鹏
责任校对：胡伟民
责任印制：丛怀宇

出版发行：清华大学出版社
网　　址：https://www.tup.com.cn，https://www.wqxuetang.com
地　　址：北京清华大学学研大厦A座　　邮　编：100084
社　总　机：010-83470000　　邮　购：010-62786544
投稿与读者服务：010-62776969，c-service@tup.tsinghua.edu.cn
质　量　反　馈：010-62772015，zhiliang@tup.tsinghua.edu.cn
课　件　下　载：https://www.tup.com.cn，010-83470236
印 装 者：北京瑞禾彩色印刷有限公司
经　　销：全国新华书店
开　　本：185mm×260mm　　印　张：15.5　　字　数：390千字
版　　次：2025年4月第1版　　印　次：2025年4月第1次印刷
定　　价：79.80元

产品编号：087126-01

前言

　　PyTorch 是 Python 实现的一个深度学习框架，也是一个使用 GPU 和 CPU 进行深度学习的优化张量库。目前，PyTorch 的主要维护者是 Facebook AI Research（FAIR），它是一个由 Facebook 和其他研究机构共同开发的开源项目。本书介绍了使用 PyTorch 开发深度学习应用。

　　本书共 5 章，第 1 章为深度学习快速入门；第 2 章为 Python 技术基础；第 3 章为 PyTorch 中的深度学习；第 4 章为 PyTorch 开发深度学习应用；第 5 章为分布式 PyTorch。

　　书中难免存在错误和疏漏，在此恳请广大读者朋友不吝指正。

　　感谢早期合著者、合作伙伴、员工、学员、读者的支持，给我们提供了良好的工作基础。就像玻璃容器中的水培植物一样，这是一个持久可用的成长基础。技术的融合与创新无止境。欢迎一起探索。

罗刚

2024 年 9 月

目 录

第1章 深度学习快速入门 ··· 1
- 1.1 各种深度学习应用 ·· 1
- 1.2 准备开发环境 ·· 2
  - 1.2.1 Linux 基础 ·· 2
  - 1.2.2 Micro 编辑器 ··· 4
  - 1.2.3 在 Linux 系统中安装 Python ······························ 5
  - 1.2.4 选择 Python 版本 ··· 5
  - 1.2.5 在 Windows 系统中安装 Python ·························· 6
- 1.3 体验 PyTorch ·· 8
  - 1.3.1 安装 PyTorch ··· 8
  - 1.3.2 PyTorch 中的计算图 ··· 8
  - 1.3.3 用三阶多项式拟合函数 ···································· 11
  - 1.3.4 实现手写数字识别 ·· 15
- 1.4 本章小结 ··· 22

第2章 Python 技术基础 ··· 23
- 2.1 变量 ··· 23
- 2.2 注释 ··· 23
- 2.3 简单数据类型 ·· 23
  - 2.3.1 数值 ·· 24
  - 2.3.2 字符串 ··· 26
  - 2.3.3 数组 ·· 28
- 2.4 字面值 ·· 28
- 2.5 控制流 ·· 29
  - 2.5.1 条件语句 ·· 29
  - 2.5.2 循环语句 ·· 30
- 2.6 列表 ··· 31
- 2.7 元组 ··· 34

2.8	集合	36
2.9	字典	36
2.10	位数组	37
2.11	模块	38
2.12	函数	39
2.13	print 函数	41
2.14	正则表达式	43
2.15	文件操作	45
	2.15.1 读写文件	46
	2.15.2 重命名文件	47
	2.15.3 遍历文件	47
2.16	with 语句	48
2.17	使用 pickle 模块序列化对象	49
2.18	面向对象编程	49
2.19	命令行参数	51
2.20	数据库	52
2.21	JSON 格式	53
2.22	日志记录	54
2.23	异常处理	55
2.24	本章小结	55

第 3 章 PyTorch 中的深度学习 57

3.1	神经网络基础	57
	3.1.1 实现深度前馈网络	57
	3.1.2 计算过程	59
3.2	卷积神经网络	64
3.3	PyTorch 基础知识	70
	3.3.1 创建张量	70
	3.3.2 随机张量	72
	3.3.3 零和一	73
	3.3.4 范围张量	73
	3.3.5 张量数据类型	74
	3.3.6 从张量获取信息	74
	3.3.7 操纵张量	75

		3.3.8	深度学习中最常见的错误之一（形状错误）	77

		3.3.8 深度学习中最常见的错误之一（形状错误）	77
		3.3.9 求最小值、最大值、平均值、总和等	80
		3.3.10 最大值、最小值的所处位置	81
		3.3.11 更改张量数据类型	81
		3.3.12 重塑、堆叠、压缩和解压	82
		3.3.13 索引（从张量中选择数据）	84
		3.3.14 PyTorch 张量和 NumPy	85
		3.3.15 再现性（试图从随机中提取随机性）	86
3.4	transformer 架构			88
		3.4.1 编码器	89
		3.4.2 解码器	94
		3.4.3 生成概率分布	98
3.5	为 PyTorch 模型提供服务			98
3.6	本章小结			102

第 4 章 PyTorch 开发深度学习应用 104

4.1	文本分类		104
	4.1.1	准备数据集	105
	4.1.2	定义网络	106
	4.1.3	训练网络	107
4.2	开发聊天机器人		109
4.3	用 Wav2Vec 2.0 进行语音识别		134
4.4	机器翻译		138
4.5	本章小结		145

第 5 章 分布式 PyTorch 146

5.1	PyTorch 分布式概述		146
	5.1.1	数据并行训练	147
	5.1.2	基于 RPC 的分布式训练	148
5.2	数据并行		148
5.3	单机模型并行最佳实践		152
5.4	分布式数据并行入门		158
5.5	用 PyTorch 编写分布式应用程序		164
5.6	完全分片数据并行入门		170
5.7	基于完全分片数据并行的高级模型训练		178

5.8	分布式 RPC 框架入门	189
	5.8.1 使用 RPC 和 RRef 的分布式强化学习	190
	5.8.2 使用分布式 Autograd 和分布式优化器的分布式 RNN	196
5.9	使用分布式 RPC 框架实现参数服务器	200
5.10	基于 RPC 的分布式流水线并行	210
	5.10.1 步骤 1：分区 ResNet50 模型	210
	5.10.2 步骤 2：将 ResNet50 模型分片拼接到一个模块中	212
	5.10.3 步骤 3：定义训练循环	214
	5.10.4 步骤 4：启动 RPC 进程	215
5.11	使用异步执行实现批量 RPC 处理	216
	5.11.1 批量更新参数服务器	216
	5.11.2 批量处理 CartPole 求解器	219
5.12	分布式数据并行与分布式 RPC 框架的结合	225
5.13	使用流水线并行性训练 transformer 模型	231
5.14	本章小结	237

第 1 章 深度学习快速入门

来源于人工神经网络的深度学习技术随着可用于机器学习数据的积累而迅速发展。深度学习方法可以从大量的训练数据中自动学习出实例的特征，在语音对话、图像识别、自然语言处理等领域得到了广泛的应用。PyTorch 是一个流行的深度学习框架。

1.1 各种深度学习应用

深度学习在很多领域都有广泛的应用，以下列举一些主要应用领域。
- 计算机视觉。深度学习在图像分类、目标检测、人脸识别、图像生成等方面应用广泛，如 Google 的 Inception 网络、Facebook 的 ResNet 等。
- 自然语言处理。深度学习在自然语言处理领域应用广泛，如机器翻译、情感分析、文本分类、语音识别、问答系统等，典型应用有 Google 的 BERT、OpenAI 的 GPT 等。
- 语音识别。深度学习在语音识别领域应用广泛，如语音转文字、语音合成等，如百度的 DeepSpeech2 等。
- 推荐系统。深度学习在推荐系统领域应用广泛，如电商推荐、社交媒体推荐等，如 Netflix 的推荐系统、Amazon 的推荐系统等。
- 医疗健康。深度学习在医疗健康领域应用广泛，如医学影像分析、疾病预测、药物发现等，如 Google 的 DeepVariant、IBM 的 Watson Health 等。
- 金融。深度学习在金融领域应用广泛，如风险管理、信用评估、股票预测等，如汇丰银行的风险管理系统、花旗银行的股票预测系统等。
- 自动驾驶。深度学习在自动驾驶领域应用广泛，如图像识别、目标检测、行驶路径规划等，如 Waymo 的自动驾驶系统、Tesla 的自动驾驶系统等。

这只是深度学习应用领域的冰山一角，随着深度学习技术的不断发展，未来还将涉及更多领域。

1.2 准备开发环境

当前，很多语音识别应用是在 Linux 操作系统下开发的。Linux 是围绕 Linux 内核构建的免费和开源软件操作系统系列。Linux 来源于 UNIX，Linux 是 UNIX 操作系统的开放源代码实现。通常，Linux 以桌面和服务器使用的称为 Linux 发行版的形式打包。Linux 有一些常用的发行版，如 CentOS 和 Ubuntu 等。Ubuntu 是由 Canonical 公司开发的基于 Debian 的开源 Linux 操作系统。CentOS 是 Red Hat Enterprise Linux 的免费克隆版。本节首先介绍在 Ubuntu 和 CentOS 下安装 Python，然后介绍在 Linux 下开发 Python 应用的编辑器 Micro。

在 Windows 下，可以使用 Notepad++ 这样的文本编辑器编写 Python 代码，也可以使用 PyCharm 或者 PyDev 集成开发环境编写代码。

1.2.1 Linux 基础

有些深度学习系统运行在 Linux 服务器中，为了远程登录 Linux 服务器，可以安装 KiTTY（https://www.fosshub.com/KiTTY.html）。在 KiTTY 的配置界面输入 IP 地址、用户名和密码后，登录 Linux 服务器。如果是用 root 账户登录，则终端提示符是 #，否则，终端提示符是 $。

查看 Ubuntu 操作系统版本号的代码如下：

```
$cat /etc/issue
Ubuntu 18.04 LTS \n \l
```

或者

```
$ lsb_release -r
Release:        18.04
```

获取 Ubuntu 的代号的代码如下：

```
$ lsb_release -c
Codename:       bionic
```

ls 命令用于列出当前目录下的文件。有的命令比较长，为了快速输入，可以用 Tab 键补全命令。history 命令用于显示历史命令。用上方向键可以选择最近运行过的命令再次执行。

可以使用支持 SSH 协议的终端仿真程序 SecureCRT 连接到远程 Linux 服务器。因为它可以保存登录秘密，所以比较方便。除了 SecureCRT，还可以使用开源软件 PuTTY（http://www.chiark.greenend.org.uk/~sgtatham/putty），以及可以保存登录密码的 PuTTY Connection Manager。

在终端启动的进程断开连接后会停止运行。为了让进程继续运行，可以使用 nohup 命令。

如果需要安装软件，可以下载对应的 RPM 安装包，然后使用 RPM 命令安装。但操作系统对应的 RPM 安装包找起来往往比较麻烦。一个软件包可能依赖其他的软件包，为了安装一个软件，需要下载几个它所依赖的软件包。

为了简化安装操作，可以使用黄狗升级管理器（Yellowdog Updater Modified），一般简称 yum。yum 会自动计算出程序之间的相互关联性，并且计算出完成软件包的安装需要哪些步骤。这样，在安装软件时，不会再被那些关联性问题所困扰。

yum 软件包管理器自动从网络下载并安装软件。yum 类似于 360 软件管家，但是不会有商业倾向的推销软件。例如，安装支持 wget 和 rzsz 命令的软件的代码如下：

```
#yum install wget
#yum install lrzsz
```

Windows 格式文本文件的换行符为 \r\n，而 Linux 文件的换行符为 \n。

dos2unix 是将 Windows 格式文件转换为 Linux 格式的实用命令。dos2unix 命令其实就是将文件中的 \r\n 转换为 \n。

开发深度学习系统的过程中，可能会用到大量的数据文件。如果需要在 Linux 操作系统中维护同一个文件的两份或多份副本，除了保存多份单独的物理文件副本，还可以采用保存一份物理文件副本和多个虚拟副本的方法。这种虚拟的副本就称为链接。链接是目录中指向文件真实位置的占位符。在 Linux 中有两种不同类型的文件链接：符号链接和硬链接。

符号链接就是一个实实在在的文件，它指向存放在虚拟目录结构中某个地方的另一个文件。这两个通过符号链接在一起的文件，彼此的内容并不相同。

如果现有空间不够用，可以增加存储设备后扩容。首先用 lsblk 命令查看现有空间情况。在一个 Linux 账号中显示如下：

```
[root@localhost ~]# lsblk
NAME    MAJ:MIN RM  SIZE RO TYPE MOUNTPOINT
sr0      11:0    1 1024M  0 rom
xvda    202:0    0  100G  0 disk
├─xvda1 202:1    0    8G  0 part [SWAP]
└─xvda2 202:2    0   32G  0 part /mnt
xvde    202:64   0 1000G  0 disk
```

创建一个要扩展的目录，代码如下：

```
# mkdir /ext
```

加载文件系统到这个目录下，代码如下：

```
# mount /dev/xvde /ext
```

确认加载成功，代码如下：

```
[root@localhost ~]# df -m
Filesystem     1M-blocks  Used  Available  Use%  Mounted on
/dev/xvda2         32752 32496        256  100%  /
devtmpfs           32108     0      32108    0%  /dev
tmpfs              32020     1      32020    1%  /dev/shm
tmpfs              32020   186      31835    1%  /run
tmpfs              32020     0      32020    0%  /sys/fs/cgroup
tmpfs               6374     1       6374    1%  /run/user/0
/dev/xvde        1007801    77     956509    1%  /ext
```

对于大的文件，可以使用 wget 命令在后台下载，代码如下：

```
#wget -bc <path>
```

这里的参数 b 表示在后台运行，参数 c 表示支持断点续传。

可以在 Windows 操作系统中编辑文本文件，然后使用 perl 命令把 Windows 操作系统中的文本文件转换成 Linux 可以识别的格式：

```
#perl -p -e 's/\r$//' < winfile.txt > unixfile.txt
```

在 Ubuntu 操作系统中安装 Python3，代码如下：

```
#apt install python3
```

Ubuntu 系统上默认的 root 密码是随机生成的，而 root 权限是通过 'sudo' 命令授予的。可以在终端输入命令 sudo passwd，然后输入当前用户的密码，按 Enter 键，终端会提示输入新的密码并确认，此时的密码就是 root 的新密码。修改成功后，输入命令 su root，再输入新的密码即可。

1.2.2 Micro 编辑器

为了方便在服务器端开发 Python\Perl\Shell\C++ 相关应用，可以使用 Micro（https://github.com/zyedidia/micro）这样的终端文本编辑器。

可以在 /home/soft/micro 目录下运行，代码如下：

```
# curl https://getmic.ro | bash
```

设置成在任意路径均可运行 Micro，代码如下：

```
#cd /usr/bin
#sudo ln -s /home/soft/micro/micro micro
```

或者编辑 /etc/profile 文件，增加 micro 所在的路径到 PATH 环境变量 /home/soft/micro，代码如下：

```
# ./micro /etc/profile
```

增加如下行:

```
export PATH=/home/soft/micro:$PATH
```

可以使用它编辑配置文件,代码如下:

```
#./micro   test.py
```

输入:

```
print("Hello World")
```

保存文件后,按 Ctrl+Q 组合键退出。

1.2.3 在 Linux 系统中安装 Python

检查 Python 3 是否已经正确安装,并确认其版本号,代码如下:

```
# python3 -V
Python 3.4.5
```

检查 Python 3 所在的路径,代码如下:

```
# which python3
/usr/bin/python3
```

如果使用 CentOS,可以使用 yum 命令安装 Python 3。首先查找可供安装的 Python 版本,代码如下:

```
# yum search python3
```

然后安装想要的版本,代码如下:

```
# yum install python36
```

如果使用 Ubuntu 操作系统,首先运行以下命令更新软件包列表并将所有系统软件升级到可用的最新版本。

```
# sudo apt-get update && sudo apt-get -y upgrade
```

然后安装 Pip 包管理系统,代码如下:

```
#sudo apt-get install python3-pip
```

1.2.4 选择 Python 版本

Linux 系统中有可能同时存在多个可用的 Python 版本。每个 Python 版本都对应一个可执行二进制文件。可以使用 ls 命令查看系统中有哪些 Python 的二进制文件可供使用，代码如下：

```
$ ls /usr/bin/python*
```

python 命令执行 Python 2。可以使用 python3 命令执行 Python 3。如何使用 python 命令执行 Python 3？

一种简单又安全的方法是使用别名。将如下命令放入 ~/.bashrc 或 ~/.bash_aliases 文件中：

```
alias python=python3
```

最好在终端使用 'python3' 命令，在 Python 3.x 文件中使用 shebang 行 '#!/usr/bin/env python3'。

1.2.5 在 Windows 系统中安装 Python

在图形化用户界面出现之前，人们就是用命令行来操作计算机的。Windows 命令行是通过 Windows 系统目录下的 cmd.exe 执行的。执行这个程序最直接的方式是找到这个程序，然后双击。因为 cmd.exe 并没有一个桌面的快捷方式，所以，这样太麻烦。

选择"开始"→"运行"命令，或者按窗口 +R 组合键，弹出"运行"对话框。输入 cmd 后单击"确定"按钮，出现命令提示窗口。因为能够能通过这个黑屏的窗口直接输入命令来控制计算机，所以也称为控制台窗口。

开始的路径往往是 C:\Users\Administrator。Windows 命令行也有个当前目录的概念，C:\Users\Administrator 就是当前路径。

可以用 cd 命令改变当前路径，例如，改变到 C:\Python\Python37 路径，代码如下：

```
C:\Users\Administrator>cd C:\Python\Python37
```

系统约定从指定的路径找可执行文件。这个路径通过 PATH 环境变量指定。环境变量是一个"变量名 = 变量值"的对应关系，每个变量都有一个或者多个值与之对应。如果是多个值，则这些值之间用分号分开。例如，PATH 环境变量可能对应如下值："C:\Windows\system32;C:\Windows"，表示 Windows 会从 C:\Windows\system32 和 C:\Windows 两个路径找可执行文件。

设置或者修改环境变量的具体操作步骤如下：

（1）在 Windows 桌面上右击"此电脑"图标，在弹出的快捷菜单中选择"属性"命令。

（2）在"系统属性"界面中单击"高级系统设置"按钮。

（3）在"高级选项"模块中单击"环境变量"按钮。

（4）弹出"环境变量"对话框，可以看到"用户变量"和"系统变量"两个模块。如果想修改或创建系统环境变量，可单击相应的"新建"按钮或选择已有的变量后单击"编辑"按钮进行修改。

（5）在"新建"对话框中输入变量的名称及变量值。变量值通常是文件路径，也可以是一个固定的值。

可以用如上步骤设置环境变量 PATH 的值。

重新启动命令行才能让环境变量的设置生效。为了检查环境变量是否设置正确，可以在命令行中显示指定环境变量的值。这时需要用到 echo 命令。echo 命令用来显示一段文字，例如：

```
C:\Users\Administrator>echo Hello
```

执行上述命令，将在命令行输出 Hello。

如果要引用环境变量的值，可以用两个百分号把变量名包围起来，即"% 变量名 %"，代码如下：

```
C:\Users\Administrator>echo %PATH%
```

假设把 Python 安装在 D:\Python\Python37 目录下，则可以在计算机属性中手工设置 PATH 环境变量，然后检查环境变量的值，代码如下：

```
C:\Users\Administrator.PC-201909301458>echo %PATH%
C:\Windows\system32;C:\Windows;C:\Windows\System32\Wbem;C:\Windows\System32\WindowsPowerShell\v1.0\;D:\apache-maven-3.5.2\bin;D:\Python\Python37
```

如下命令用于检查 Python 是否正确安装，以及所使用的版本号：

```
>python --version
```

或者用 where 命令检查系统是否已经安装了 Python。

```
C:\Users\Administrator>where python
C:\Python27\python.exe
D:\cygwin64\bin\python
D:\Programs\Python\Python37\python.exe
```

如果没有安装 Python，则可以使用 Chocolatey（https://chocolatey.org）安装 Python3 代码如下：

```
>choco install python3
```

或者访问 Python 的官方网站（https://www.python.org/）并下载最新版本。当前最新版本是 Python 3.12。

```
wget https://www.python.org/ftp/python/3.12.2/python-3.12.2-amd64.exe
```

确保选择"添加 python.exe 到 PATH"选项，然后单击"立即安装"按钮。

1.3 体验 PyTorch

本节以手写数字识别这个经典问题来体验 PyTorch。

1.3.1 安装 PyTorch

本节介绍在 Windows 操作系统中安装 PyTorch。

要仅为在 CPU 上使用而安装 PyTorch 的当前版本：

```
pip install torch torchvision torchaudio
```

如果要使用支持 CUDA 的 GPU 卡，则安装 PyTorch 的 GPU 版本。如果有 CUDA 11.8 支持的 GPU，可以运行以下命令：

```
pip install torch torchvision torchaudio --index-url https://download.pytorch.org/whl/cu118
```

也可以先下载相应的安装包，然后从本地安装 whl 文件。whl 文件下载地址是 https://download.pytorch.org/whl/torch_stable.html。安装过程如下：

```
wget -c https://download.pytorch.org/whl/cpu/torch-2.2.1%2Bcpu-cp312-cp312-win_amd64.whl
pip install torch-2.2.1+cpu-cp312-cp312-win_amd64.whl
```

在交互式环境测试 PyTorch。
首先，输入以下代码导入 PyTorch 包。

```
>>> import torch
```

然后按 Enter 键。

定义一个由 0 组成的向量。现在，把向量想象成一个数字集合或一个数字列表。接下来，使用 Tensor() 创建一个包含 3 个数字的列表的向量。这是一个三维矢量，它是三维空间中的一个箭头。

运行以下 Python 代码：

```
>>> torch.Tensor([0, 0, 0])
tensor([0., 0., 0.])
```

这表明 PyTorch 安装成功。按 Ctrl+D 组合键退出 Python 交互式控制台。

1.3.2 PyTorch 中的计算图

PyTorch 中的计算图是一种图形结构，用于表示计算过程的依赖关系。在 PyTorch 中，计算图是通过自动微分（Autograd）实现的。Autograd 跟踪所有在张量上执行的操作，并构建计算图，同时计算操作的梯度，以便在反向传播过程中更新参数。

PyTorch 张量表示计算图中的一个节点。如果 x 是一个张量，其 x.requires_grad=True，那么 x.grad 是另一个张量相对于某个标量值保持 x 的梯度。

下面是一个使用 PyTorch 构建计算图的示例代码。

```
import torch

# 定义输入张量
x = torch.tensor([[1.0, 2.0], [3.0, 4.0]], requires_grad=True)
# 定义权重张量
w = torch.tensor([[5.0, 6.0], [7.0, 8.0]], requires_grad=True)
# 定义偏置张量
b = torch.tensor([[9.0, 10.0]], requires_grad=True)

# 定义计算图
y = torch.matmul(x, w) + b
z = torch.sum(y)

# 计算梯度
z.backward()

# 输出梯度
print(x.grad)
print(w.grad)
print(b.grad)
```

在 PyTorch 中，每个张量都有一个 grad_fn 属性，该属性记录了创建该张量的操作，也称为梯度函数。梯度函数是用于计算张量的梯度（或导数）的函数，其根据链式法则将梯度向后传播到计算图中的先前计算节点，例如：

```
import torch

# 创建一个指定形状的张量，并将其所有元素初始化为1
x = torch.ones(3, 2, requires_grad=True)
print(x)
print(x.grad_fn)      # 输出 :None
```

当在计算图上执行反向传播时，PyTorch 使用每个张量的 grad_fn 属性来构造反向传播图。这样，每个张量都知道如何将其梯度向后传播到先前的操作。

torch.no_grad() 是一个上下文管理器，它将禁用上下文中的所有梯度计算，例如：

```
import torch
import math
```

```python
dtype = torch.float
device = torch.device("cpu")

# Create tensors to hold input and outputs
# As we don't need to compute gradients with respect to these Tensors, we can set
# requires_grad = False. This is also the default setting.
x = torch.linspace(-math.pi, math.pi, 2000)
y = torch.sin(x)

# Create random tensors for weights. For these Tensors, we require gradients,
# therefore, we can set requires_grad = True
a = torch.randn((), device = device, dtype = dtype, requires_grad=True)
b = torch.randn((), device = device, dtype = dtype, requires_grad=True)
c = torch.randn((), device = device, dtype = dtype, requires_grad=True)
d = torch.randn((), device = device, dtype = dtype, requires_grad=True)

learning_rate = 1e-6

# Forward pass: we compute predicted y using operations on Tensors.
y_pred = a + b * x + c * x ** 2 + d * x ** 3

# Compute and print loss using operations on Tensors.
# Now loss is a Tensor of shape (1,)
# loss.item() gets the scalar value held in the loss.
loss = (y_pred - y).pow(2).sum()

# Use autograd to compute the backward pass. This call will compute the
# gradient of loss with respect to all Tensors with requires_grad=True.
# After this call a.grad, b.grad. c.grad and d.grad will be Tensors holding
# the gradient of the loss with respect to a, b, c, d respectively.
loss.backward()

# Manually update weights using gradient descent. Wrap in torch.no_grad()
# because weights have requires_grad=True, but we don't need to track this
# in autograd.
with torch.no_grad():
    a -= learning_rate * a.grad
    b -= learning_rate * b.grad
    c -= learning_rate * c.grad
    d -= learning_rate * d.grad

    # Manually zero the gradients after updating weights
    a.grad = None
    b.grad = None
    c.grad = None
```

```
d.grad = None
```

1.3.3　用三阶多项式拟合函数

本节将拟合 y=sin(x) 从 −π 到 π 上的三阶多项式。多项式有 4 个参数，将使用梯度下降来通过最小化预测输出和真实输出之间的欧几里得距离来拟合随机数据。

下面将介绍如下 3 种拟合多项式的方法。
- 使用 NumPy 实现多项式拟合并使用 NumPy 操作手动实现正向和反向通过。
- 利用 PyTorch 张量的概念。
- 使用 PyTorch 中的 Autograd 软件包，该软件包使用自动微分来自动计算后向传播。

NumPy 是一个很好的科学计算工具，但对于深度学习来说不是很方便，因为它对梯度或计算图一无所知。然而，将三阶多项式拟合到正弦函数是非常容易的。

程序用到了绘图库 Matplotlib。Matplotlib 是 Python 编程语言及其数值数学扩展 NumPy 的绘图库。安装 Matplotlib 的代码如下：

```
pip install matplotlib
```

拟合多项式实现代码如下：

```
import numpy as np
import math
import matplotlib.pyplot as plt

x = np.linspace(-math.pi, math.pi, 2000)
y = np.sin(x)

# We randomly initialize weights
a = np.random.randn()
b = np.random.randn()
c = np.random.randn()
d = np.random.randn()

# print randomly initialized weights
print(f'a = {a}, b = {b}, c = {c}, d = {d}')

# learning rate
lr = 1e-6

for i in range(5000):
    # y = a + bx + cx^2 + dx^3
```

```python
    y_pred = a + b*x + c*x ** 2 + d*x ** 3

    # Compute and print loss
    loss = np.square(y_pred -y).sum()
    if i%100 == 0:
      print(i,loss)

    # Backprop to compute the gradients of a, b, c, d with respect to loss
    #dL/da = (dL/dy_pred) * (dy_pred/da)
    #dL/db = (dL/dy_pred) * (dy_pred/db)
    #dL/dc = (dL/dy_pred) * (dy_pred/dc)
    #dL/dd = (dL/dy_pred) * (dy_pred/dd)

    grad_y_pred = 2.0 * (y_pred-y)
    grad_a = grad_y_pred.sum()
    grad_b = (grad_y_pred * x).sum()
    grad_c = (grad_y_pred * x ** 2).sum()
    grad_d = (grad_y_pred * x ** 3).sum()

    # Update Weights
    a -= lr * grad_a
    b -= lr * grad_b
    c -= lr * grad_c
    d -= lr * grad_d

plt.plot(x,y,label = 'y = sin(x)', c = 'b')
plt.plot(x, y_pred, label = 'y = a + bx + cx^2 + dx^3', c = 'r',linestyle = 'dashed')
plt.xlabel('x')
plt.ylabel('y')
plt.ylim([-2,2])
plt.legend()
plt.show()

print(f'Result: y = {a} + {b} x + {c} x^2 + {d} x^3')
```

使用 NumPy 拟合三阶多项式很容易。但现代深度神经网络呢？不幸的是，NumPy 无法利用 GPU 来加速其数值计算。这就是 PyTorch 张量的有用之处。张量是一个 n 维数组，可以跟踪梯度和计算图。要在 GPU 上运行 PyTorch 张量，只需要指定正确的设备。但就目前而言，我们将坚持使用 CPU。

看看如何使用 PyTorch 张量来完成我们的任务。

```python
import torch
import math
```

```python
dtype = torch.float
device = torch.device("cpu")
#device = torch.device("cuda:0") # Uncomment this if GPU is available.

# Create random input and data
x = torch.linspace(-math.pi, math.pi, 2000, device=device, dtype=dtype)
y = torch.sin(x)

# Randomly initialize weights
a = torch.randn((), device=device, dtype=dtype)
b = torch.randn((), device=device, dtype=dtype)
c = torch.randn((), device=device, dtype=dtype)
d = torch.randn((), device=device, dtype=dtype)

learning_rate = 1e-6

for t in range(5000):
    # Forward pass: compute predicted y
    y_pred = a + b * x + c * x ** 2 + d * x ** 3

    # Compute and print loss
    loss = (y_pred - y).pow(2).sum().item()
    if t % 100 == 99:
        print(t, loss)

    # Backprop to compute gradients of a, b, c, d with respect to loss
    grad_y_pred = 2.0 * (y_pred - y)
    grad_a = grad_y_pred.sum()
    grad_b = (grad_y_pred * x).sum()
    grad_c = (grad_y_pred * x ** 2).sum()
    grad_d = (grad_y_pred * x ** 3).sum()

    # Update weights using gradient descent
    a -= learning_rate * grad_a
    b -= learning_rate * grad_b
    c -= learning_rate * grad_c
    d -= learning_rate * grad_d

plt.plot(x,y,label = 'y = sin(x)', c = 'b')
plt.plot(x, y_pred, label = 'y = a + bx + cx^2 + dx^3', c = 'r',linestyle = 'dashed')
plt.xlabel('x')
plt.ylabel('y')
```

```python
plt.ylim([-2,2])
plt.legend()
plt.show()
print(f'Result: y = {a.item()} + {b.item()} x + {c.item()} x^2 + {d.item()} x^3')
```

我们在上面看到了张量也可以用于将三阶多项式拟合到 sin 函数。然而，我们不得不手动包含前向和后向传播。这对于拟合多项式这样的简单任务来说并不难，但对于深度神经网络来说，可能会变得非常混乱。幸运的是，PyTorch 的 Autograd 软件包可以用来自动计算向后传播，实现代码如下：

```python
import torch
import math
import matplotlib.pyplot as plt

dtype = torch.float
device = torch.device("cpu")

# Create tensors to hold input and outputs
# As we don't need to compute gradients with respect to these Tensors, we can set
requires_grad = False. This is also the default setting.

x = torch.linspace(-math.pi, math.pi, 2000)
y = torch.sin(x)

# Create random tensors for weights. For these Tensors, we require gradients,
therefore, we can set requires_grad = True

a = torch.randn((), device=device, dtype=dtype, requires_grad=True)
b = torch.randn((), device=device, dtype=dtype, requires_grad=True)
c = torch.randn((), device=device, dtype=dtype, requires_grad=True)
d = torch.randn((), device=device, dtype=dtype, requires_grad=True)

learning_rate = 1e-6

for t in range(5000):
    # Forward pass: we compute predicted y using operations on Tensors.
    y_pred = a + b * x + c * x ** 2 + d * x ** 3

    # Compute and print loss using operations on Tensors.
    # Now loss is a Tensor of shape (1,)
    # loss.item() gets the scalar value held in the loss.
    loss = (y_pred - y).pow(2).sum()
    if t % 100 == 99:
        print(t, loss.item())
```

```
# Use autograd to compute the backward pass. This call will compute the
# gradient of loss with respect to all Tensors with requires_grad=True.
# After this call a.grad, b.grad. c.grad and d.grad will be Tensors holding
# the gradient of the loss with respect to a, b, c, d respectively.
loss.backward()

# Manually update weights using gradient descent. Wrap in torch.no_grad()
# because weights have requires_grad=True, but we don't need to track this
# in autograd.
with torch.no_grad():
    a -= learning_rate * a.grad
    b -= learning_rate * b.grad
    c -= learning_rate * c.grad
    d -= learning_rate * d.grad

    # Manually zero the gradients after updating weights
    a.grad = None
    b.grad = None
    c.grad = None
    d.grad = None

plt.plot(x, y, label='y = sin(x)', c='b')
# We need to use tensor.detach().numpy() to convert our tensor into numpy array for plotting
plt.plot(x, y_pred.detach().numpy(), label='y = a + bx + cx^2 + dx^3', c='r', linestyle='dashed')
plt.xlabel('x')
plt.ylabel('y')
plt.ylim([-2, 2])
plt.legend()
plt.show()
print(f'Result: y = {a.item()} + {b.item()} x + {c.item()} x^2 + {d.item()} x^3')
```

1.3.4 实现手写数字识别

首先建立第一个神经网络并对其进行训练。PyTorch 中的子库 torch.nn 用于神经网络操作，子库 torch.optim 用于神经网络优化器。优化器是一种函数或算法，用于修改神经网络的属性，如权重和学习率。

这里将使用以下 5 个步骤来构建和训练模型：构建计算图→设置优化器→设置标准→设置数据→训练模型。

首先使用可管理的数据集构建一个小型模型。

创建一个新文件 step_2_helloworld.py。

编写一个简短的 18 行代码片段，用于训练一个小模型。首先导入几个 PyTorch 模块，代码如下：

```
import torch
import torch.nn as nn
import torch.optim as optim
```

在这里，将 PyTorch 库别名分为以下几个常用的快捷方式：
- torch 包含所有 PyTorch 实用程序。然而，常规 PyTorch 代码包括一些额外的导入。在这里遵循相同的约定，这样就可以在线理解 PyTorch 教程和随机代码片段。
- torch.nn 包含用于构建神经网络的实用程序。这通常表示为 nn。
- torch.optim 包含训练实用程序。这通常被表示为 optim。

接下来，定义神经网络、训练使用程序和数据集，代码如下：

```
...
net = nn.Linear(1, 1)  # 1. Build a computation graph (a line!)
optimizer = optim.SGD(net.parameters(), lr=0.1)  # 2. Setup optimizers
criterion = nn.MSELoss()  # 3. Setup criterion
x, target = torch.randn((1,)), torch.tensor([0.])  # 4. Setup data
...
```

上述代码定义了任何深度学习训练脚本的几个必要部分。
- net = ... 定义了"神经网络"。在这种情况下，模型是一条形式为 y=m*x 的线；参数 nn.Linear(1, 1) 是直线的斜率。此模型参数 nn.Linear(1, 1) 将在训练期间更新。注意，torch.nn 包括许多深度学习运算，如这里使用的全连接层（nn.Lineral）和卷积层（nn.Conv2d）。
- optimizer = ... 定义优化器。这个优化器决定了神经网络将如何学习。torch.optim（别名为 optim）包括许多可以使用的此类优化器。
- criterion = ... 定义了损失。简而言之，损失定义了模型试图最小化的内容。对于直线的基本模型，目标是最小化直线的预测 y 值与训练集中的实际 y 值之间的差异。注意，torch.nn 包括许多其他可以使用的损失函数。
- x, target = ... 定义了数据集。现在，数据集只是一个坐标，一个 x 值和一个 y 值。在这里，torch 包本身提供 tensor 来创建新的张量，并提供 randn 来创建具有随机值的张量。

最后，通过在数据集上迭代 10 次来训练模型，每次都要调整模型的参数，代码如下：

```
...
# 5. Train the model
for i in range(10):
    output = net(x)
```

```
    loss = criterion(output, target)
    print(round(loss.item(), 2))

    net.zero_grad()
    loss.backward()
    optimizer.step()
```

总体目标是通过调整线的斜率来最大限度地减少损失。为了实现这一点,该训练代码实现了一种名为梯度下降的算法。

要找到损耗最低的最佳模型,可执行以下操作:

① 使用 net = nn.Linear(1, 1) 初始化一个随机模型。
② 在 for i in range(10) 循环中开始训练。
③ 每一步的方向由梯度决定。
④ 用 optimizer = ... 中的 lr=0.1 来指定步长。这决定了每个步骤可以有多大。

上述代码中的最后 3 行也很重要。

- net.zero_grad 清除上一步迭代中可能剩余的所有梯度。
- loss.backward 计算新的梯度。
- optimizer.step 使用这些梯度来采取措施。

"hello world"神经网络到此结束,保存并关闭文件即可。

运行脚本 step_2_helloworld.py。脚本将输出以下内容:

```
Output
0.33
0.19
0.11
0.07
0.04
0.02
0.01
0.01
0.0
0.0
```

注意,损失不断减少,这表明模型正在学习。使用 PyTorch 时,还需要注意如下两个实现细节。

- PyTorch 使用 torch.Tensor 保存所有数据和参数。这里,torch.randn 生成一个具有随机值的张量,该张量具有所提供的形状。例如,torch.randn((1, 2)) 创建一个 1×2 张量或二维行向量。
- PyTorch 支持多种优化器。这里使用了 torch.optim.SGD,也称为随机梯度下降(SGD)。还有更多的优化器在 SGD 的基础上添加了额外的功能。也有许多损失,torch.nn.MSELoss 只是其中之一。

这就结束了在玩具数据集上的第一个模型。下一步，将使用神经网络代替这个小模型，用常用的机器学习基准代替玩具数据集。

为了更好地理解 PyTorch 的优点，下面使用包含更多神经网络操作的 torch.nn.functional 和支持许多可以开箱即用的数据集的 torchvision.datasets 构建深度神经网络。接下来，将使用预制数据集构建一个相对复杂的自定义模型。

下面将使用卷积，这是一种模式查找器。对于图像，卷积在不同级别的"意义"上寻找 2D 模式：直接应用于图像的卷积在寻找"较低级别"的特征，如边缘。然而，应用于许多其他操作的输出的卷积可能正在寻找"更高层次"的特征，如门。

现在，将通过定义一个稍微复杂一点的模型来扩展构建的第一个 PyTorch 模型。神经网络现在将包含两个卷积和一个完全连接的层，以处理图像输入。

创建一个新文件 step_3_mnist.py，遵循与之前相同的五步算法：构建计算图→设置优化器→设置标准→设置数据→训练模型。

第一，定义深层神经网络。这是可能在 MNIST 上发现的其他神经网络的精简版本。这样就可以在计算机上训练你的神经网络，代码如下：

```python
import torch
import torch.nn as nn
import torch.optim as optim
import torch.nn.functional as F

from torchvision import datasets, transforms
from torch.optim.lr_scheduler import StepLR

# 1. Build a computation graph
class Net(nn.Module):
    def __init__(self):
        super(Net, self).__init__()
        self.conv1 = nn.Conv2d(1, 32, 3, 1)
        self.conv2 = nn.Conv2d(32, 64, 3, 1)
        self.fc = nn.Linear(1024, 10)

    def forward(self, x):
        x = F.relu(self.conv1(x))
        x = F.relu(self.conv2(x))
        x = F.max_pool2d(x, 1)
        x = torch.flatten(x, 1)
        x = self.fc(x)
        output = F.log_softmax(x, dim=1)
        return output
net = Net()
...
```

在这里，定义了一个继承自 nn.Module 的神经网络类。神经网络中的所有操作（包括神经网络本身）都必须继承自 nn.Module。神经网络类的典型范例如下：

- 在构造函数中，定义网络所需的任何操作。在这种情况下，有两个卷积和一个完全连接的层。需要记住的一点是，构造函数总是以 super().__init__() 开头。PyTorch 希望在将模块（如 nn.Conv2d）分配给实例属性（self.conv1）之前初始化父类。
- 在 forward() 方法中，运行初始化的操作。这个方法决定了神经网络的结构，明确定义了神经网络将如何计算其预测。

该神经网络使用了如下 4 种操作。

- nn.Conv2d：卷积。卷积在图像中寻找图案。早期的卷积寻找像边缘这样的"低级"模式。网络中后来的卷积寻找"高级"模式，如狗的腿或耳朵。
- nn.Linear：一个完全连接的层。完全连接的层将所有输入特征与所有输出维度相关联。
- F.relu，F.max_pool2d：这些都是非线性的类型。relu() 是函数 $f(x) = \max(x, 0)$。max_pool() 在每个值块中取最大值。在这种情况下，将在整个图像中获取最大值。
- log_softmax：规范化向量中的所有值，使这些值之和为 1。

第二，像以前一样，定义优化器。这一次，将使用不同的优化器和不同的超参数设置。超参数配置训练，而训练调整模型参数。

```
...
optimizer = optim.Adadelta(net.parameters(), lr=1.)  # 2. Setup optimizer
...
```

第三，与以前不同，现在将使用不同的损失。这种损失用于分类问题，其中模型的输出是类索引。在这个特定的例子中，模型将输出包含在输入图像中的数字（可能是从 0 到 9 的任何数字）。

```
...
criterion = nn.NLLLoss()  # 3. Setup criterion
...
```

第四，建立数据。在这种情况下，将设置一个名为 MNIST 的数据集，该数据集以手写数字为特征。每个图像是一个包含手写数字的 28 像素×28 像素的小图像，目标是将每个手写数字分类为 0、1、2、…或 9。

```
...
# 4. Setup data
transform = transforms.Compose([
    transforms.Resize((8, 8)),
    transforms.ToTensor(),
```

```
        transforms.Normalize((0.1307,), (0.3081,))
])
train_dataset = datasets.MNIST(
    'data', train=True, download=True, transform=transform)
train_loader = torch.utils.data.DataLoader(train_dataset, batch_size=512)
...
```

在这里,在 transform=…中对图像进行预处理。通过调整图像的大小,将图像转换为 PyTorch 张量,并将张量归一化为平均值为 0,方差为 1。

在接下来的两行中,设置 train=True,因为这是训练数据集,并设置 download=True,以便在数据集尚未下载的情况下下载该数据集。

batch_size = 512 确定一次在多少个图像上训练网络。除非批量大得离谱(例如,数以万计),否则,更大的批量更适合进行大致更快的训练。

第五,训练模型。现在将在提供的数据集中对所有样本进行一次迭代,而不是在同一个样本上运行 10 次。通过一次遍历所有样本,以下是一个回合的训练:

```
...
# 5. Train the model
for inputs, target in train_loader:
    output = net(inputs)
    loss = criterion(output, target)
    print(round(loss.item(), 2))

    net.zero_grad()
    loss.backward()
    optimizer.step()
...
```

保存并关闭文件。

运行脚本 step_3_mnist.py。

```
python step_3_mnist.py
```

脚本将输出以下内容:

```
Output
2.31
2.18
2.03
1.78
1.52
1.35
1.3
1.35
1.07
```

```
1.0
...
0.21
0.2
0.23
0.12
0.12
0.12
```

注意,最终损失小于初始损失值的 10%。这意味着神经网络正在正确训练。

训练到此结束。然而,0.12 的损失很难解释:我们不知道 0.12 是 "好" 还是 "坏"。为了评估模型的性能,接下来要计算此分类模型的准确性。

在数据集的训练拆分上计算了损失值。但是,最好对数据集进行单独的验证拆分。可以使用此验证拆分来计算模型的准确性,但是,不能用它来训练。接下来,将设置验证数据集并在其上评估模型。这一步将使用与以前相同的 PyTorch 实用程序,包括 MNIST 数据集的 torchvision.dataset。

将 step_3_mnist.py 文件复制到 step_4_eval.py 中,打开文件 step_4_eval.py。

设置验证数据集:

```
...
train_loader = ...
val_dataset = datasets.MNIST(
    'data', train=False, download=True, transform=transform)
val_loader = torch.utils.data.DataLoader(val_dataset, batch_size=512)
...
```

在训练循环之后,在文件的末尾添加一个验证循环:

```
    ...
    optimizer.step()

correct = 0.
net.eval()
for inputs, target in val_loader:
    output = net(inputs)
    _, pred = output.max(1)
    correct += (pred == target).sum()
accuracy = correct / len(val_dataset) * 100.
print(f'{accuracy:.2f}% correct')
```

在这里,验证循环执行一些操作来计算准确性:

- 运行 net.eval() 可以确保神经网络处于评估模式并准备好进行验证。一些操作在评估模式下的运行方式与在训练模式下的不同。
- 对 val_loader 中的所有输入和标签进行迭代。

- 运行模型 net(inputs) 以获得每个类别的概率。
- 找到概率最高的类 output.max(1)。
- 计算正确分类的图像数量：pred==target 计算布尔值向量。.sum() 将这些布尔值强制转换为整数，并有效地计算真值的数量。
- correct / len(val_dataset) 最终计算正确分类的图像的百分比。

保存并关闭文件。

运行脚本 step_4_eval.py：

```
python step_4_eval.py
```

脚本将输出以下内容：

```
Output
2.31
2.21
...
0.14
0.2
89% correct
```

注意，具体损失值和最终精度可能会有所不同。

现在已经训练了第一个深度神经网络。可以通过调整训练的超参数来进行进一步的修改和改进，包括不同数量的回合、学习率和不同的优化器。

1.4 本章小结

PyTorch 是 Python 的一个深度学习框架，因其易用性和灵活性而广受欢迎。PyTorch 最初是由 Facebook AI Research 开发的，现在被许多公司和组织使用，包括谷歌、微软和优步。PyTorch 是开源的，可以在 GitHub 上使用。

2016 年 10 月，PyTorch 开始作为 Adam Paszke 的实习项目。当时，Adam Paszke 在 Torch 的核心开发者 Soumith Chintala 手下工作。Torch 是 LuaJIT 的科学计算框架。LuaJIT 是 Lua 语言的实时编译器。

PyTorch 目前由 Soumith Chintala、Gregory Chanan、Dmytro Dzhulgakov、Edward Yang 和 Nikita Shulga 维护，数百名才华横溢的个人以各种形式和方式做出了重大贡献。

PyTorch 是在经过修改的 BSD 许可证下发布的。

第 2 章　Python 技术基础

本章介绍开发深度学习应用需要的 Python 语言语法基础。

2.1　变量

在 Python 中，定义变量时是不声明类型的。变量在内部是有类型的。利用 type() 函数可以得到变量的类型。例如，在交互式环境中输入代码：

```
>>> a=3
>>> type(a)
<class 'int'>
```

表明变量 a 是整数类型。

2.2　注释

Python 脚本中用 # 表示注释。但如果 # 位于第一行开头，并且是 #!（称为 Shebang）则例外，它表示该脚本使用后面指定的解释器 /usr/bin/python3 解释执行。每个脚本程序只能在开头包含这个语句。

为了能够在源代码中添加中文注释，需要把源代码保存成 utf-8 格式，例如：

```
# -*- coding: utf-8 -*

# 从 torch 的 utils 工具箱中导入 tensorboard 模块，然后导入 SummaryWriter 类
from torch.utils.tensorboard import SummaryWriter
```

2.3　简单数据类型

本节介绍包括数值、字符串和数组在内的简单数据类型。

2.3.1 数值

Python 中有 3 种不同的数值类型：int（整数）、float（浮点数）和 complex（复数）。与 Java、C 语言中的 int 类型不同，Python 中的 int 类型是无限精度的，例如：

```
>>> i=32432444444444444444444444444444444444448797687567567657000000000000000000000000000000000000000000000000000000000000000000000564564
>>> i
32432444444444444444444444444444444444448797687567567657000000000000000000000000000000000000000000000000000000000000000000000564564
>>> type(i)
<class 'int'>
```

因为 Python 依据 IEEE 754 标准，使用二进制表示 float（浮点数），所以，存在表示精度的问题，例如：

```
>>> 0.1 == 0.1000000000000000000000001
True
```

可以使用 decimal 模块使用十进制表示完整的小数，例如：

```
>>> import decimal
>>> a = decimal.Decimal('0.1')
>>> b = decimal.Decimal('0.1000000000000000000000001')
>>> a == b
False
```

在傅里叶变换中会用到复数。复数在 Python 中是一个基本数据类型，例如：

```
>>> complex(2,3)
(2+3j)
```

复数有一些内置的访问器，例如：

```
>>> z = 2+3j
>>> z.real
2.0
>>> z.imag
3.0
>>> z.conjugate()
(2-3j)
```

如下内置函数支持复数：

```
>>> abs(3 + 4j)
5.0
>>> pow(3 + 4j, 2)
(-7+24j)
```

标准模块 cmath 具有处理复数的更多功能，例如：

```
>>> import cmath
>>> cmath.sin(2 + 3j)
(9.15449914691143-4.168906959966565j)
```

用于数值运算的算术运算符说明如表 2-1 所示。

表 2-1 用于数值运算的算术运算符说明

语　　法	数学含义	运算符名称
a+b	a+b	加
a−b	a−b	减
a*b	a×b	乘
a/b	a÷b	除
a//b	$\lfloor a \div b \rfloor$	地板除
a%b	a mod b	模
−a	-a	取负数
abs(a)	$\lvert a \rvert$	绝对值
a**b	a^b	指数
math.sqrt(a)	$\sqrt{a}$	平方根

对于"/"运算，即便分子和分母都是 int 类型的，返回的也是浮点数，例如：

```
>>> print(1/3)
0.3333333333333333
```

Python 支持不同的数字类型相加，它使用数字类型强制转换的方式来解决数字类型不一致的问题，就是说它会将一个操作数转换成与另一个操作数相同的数据类型。

如果有一个操作数是复数，则另一个操作数被转换为复数，例如：

```
>>> 3.0 + (5+6j)            # 非复数转复数
(8+6j)
```

整数转为浮点数：

```
>>> 6 + 7.0                 # 非浮点型转浮点型
13.0
```

在 Python 中，一行就是一条语句，可以使用反斜杠（\）将一条语句分为多行显示，例如：

```
>>> a = 1
>>> b = 2
>>> c = 3
>>> total = a + \
... b + \
... c
>>> total
6
```

2.3.2 字符串

使用 strip() 方法可以去掉字符串首尾的空格或者指定的字符，例如：

```
term = "   hi   ";              # 去除首尾空格
print(term.strip());
```

使用 split() 方法可以将句子分成单词。下例中，mary 是一个单一的字符串。虽然这是一个句子，但这些词语并没有表示成严格的单位。为此，需要一种不同的数据类型：字符串列表，其中每个字符串对应一个单词。使用 split() 方法可以把句子切分成单词。

```
>>> mary = 'Mary had a little lamb'
>>> mary.split()
['Mary', 'had', 'a', 'little', 'lamb']
```

split() 方法根据空格拆分 Mary，返回的结果是 Mary 中的单词列表。此列表包含 len() 函数演示的 5 个项目。对于 Mary，len() 函数返回字符串中的字符数（包括空格）。

```
>>> mwords = mary.split()
>>> mwords
['Mary', 'had', 'a', 'little', 'lamb']
>>> len(mwords)                 # mwords 中的项目数
5
>>> len(mary)                   # 字符数
22
```

空白字符包括空格（' '）、换行符（'\n'）和制表符（'\t'）等。.split() 可以分隔这些字符的任何组合序列：

```
>>> chom = ' colorless    green \n\tideas\n'
>>> print(chom)
 colorless    green
	ideas

>>> chom.split()
['colorless', 'green', 'ideas']
```

通过提供可选参数，.split('x') 可用于在特定子字符串 'x' 上拆分字符串。如果没有指定 'x'，则 .split() 只在所有空格上分割。

```
>>> mary = 'Mary had a little lamb'
>>> mary.split('a')                    # 根据 'a' 切分
['M', 'ry h', 'd ', ' little l', 'mb']
>>> hi = 'Hello mother,\nHello father.'
>>> print(hi)
Hello mother,
Hello father.
>>> hi.split()                         # 没有给出参数，则在空格上分割
['Hello', 'mother,', 'Hello', 'father.']
>>> hi.split('\n')                     # 仅在 '\n' 上分割
['Hello mother,', 'Hello father.']
```

但是如果想将一个字符串拆分成一个字符列表呢？在 Python 中，字符是长度为 1 的字符串。list() 函数可以将字符串转换为单个字母的列表，例如：

```
>>> list('hello world')
['h', 'e', 'l', 'l', 'o', ' ', 'w', 'o', 'r', 'l', 'd']
```

如果有一个单词列表，可以使用 .join() 方法将它们重新组合成一个单独的字符串。在"分隔符"字符串 'x' 上调用 join(y) 方法，'x'.join(y) 连接列表 y 中由 'x' 分隔的每个元素。下面，变量 mwords 中的 5 个单词用空格连接回句子字符串：

```
>>> mwords
['Mary', 'had', 'a', 'little', 'lamb']
>>> ' '.join(mwords)
'Mary had a little lamb'
```

也可以在空字符串 '' 上调用该方法作为分隔符。效果是列表中的元素连接在一起，元素之间没有任何内容。下面的代码将一个字符列表放回到原始字符串中。

```
>>> hi = 'hello world'
>>> hichars = list(hi)
>>> hichars
['h', 'e', 'l', 'l', 'o', ' ', 'w', 'o', 'r', 'l', 'd']
>>> ''.join(hichars)
'hello world'
```

一个字符串取子串的例子代码如下：

```
>>> x = "Hello World!"
>>> x[2:]
'llo World!'
>>> x[:2]
'He'
```

```
>>> x[:-2]
'Hello Worl'
>>> x[-2:]
'd!'
>>> x[2:-2]
'llo Worl'
```

使用 ord() 函数和 chr() 函数可以实现字符串和整数之间的互相转换，例如：

```
>>> a = 'v'
>>> i = ord(a)
>>> chr(i)
'v'
```

2.3.3 数组

可以使用 array（数组）存储同样数据类型的数值类型。通过 import array 导入 python 的数组类型，就可以使用 array 类型了。

例如：

```
from array import array
node=array('H')        # 存储无符号短整型的数组

node.append(12)
```

2.4 字面值

Python 包括如下几种类型的字面值。
- 数字：整数、浮点数、复数。
- 字符串：以单引号、双引号或者三引号定义字符串。
- 布尔值：True 和 False。
- 空值：None。

有 4 种不同的字面值集合，分别是列表字面值、元组字面值、字典字面值和集合字面值。示例代码如下：

```
fruits = ["apple", "mango", "orange"]              #列表
numbers = (1, 2, 3)                                #元组
alphabets = {'a':'apple', 'b':'ball', 'c':'cat'}   #字典
vowels = {'a', 'e', 'i' , 'o', 'u'}                #集合
```

```
print(fruits)
print(numbers)
print(alphabets)
print(vowels)
```

2.5 控制流

完成一件事情要有流程控制。例如,洗衣服分 3 个步骤:把脏衣服放进洗衣机;等洗衣机洗好衣服;晾衣服。这是顺序控制结构。

顺序执行的代码采用相同的缩进,称为一个代码块。Python 没有像 Java 或 C# 语言那样采用 {} 分隔代码块,而是采用代码缩进和冒号来区分代码之间的层次。

缩进的空白数量是可变的,但是所有代码块语句必须包含相同的缩进空白数量。Notepad++ 这样的文本编辑器支持选择多行代码后,按 Tab 键改变代码块的缩进格式。

控制流用来根据运行时情况调整语句的执行顺序。流程控制语句可以分为条件语句和循环语句。

2.5.1 条件语句

当路径不存在时,就创建它,可以使用条件语句来实现。条件语句的一般形式如下:

```
if 条件:
    语句1
    语句2...
elif 条件:
    语句1
    语句2...
else:
    语句1
    语句2...
语句x
```

例如,判断一个数是否是正数,代码如下:

```
x = -32.2;
isPositive = (x > 0);
if isPositive:
    print(x, " 是正数 ");
else:
print(x, " 不是正数 ");
```

这里的 if 复合语句,首行以关键字开始,以冒号(:)结束。

使用关系运算符和条件运算符作为判断依据。关系运算符返回一个布尔值。关系运算符如表 2-2 所示。

表 2-2 关系运算符

运算符	用法	含义
>	a > b	a 大于 b
>=	a >= b	a 大于或等于 b
<	a < b	a 小于 b
<=	a <= b	a 小于或等于 b
==	a == b	a 等于 b
!=	a != b	a 不等于 b

如果要针对多个值测试一个变量，则可以在 if 条件判断中使用一个集合，示例代码如下：

```
x = "Wild things"
y = "throttle it back"
z = "in the beginning"
if "Wild" in {x, y, z}: print (True)
```

2.5.2 循环语句

在 Python 中，可以使用 for 循环或者 while 循环实现多次重复执行一个代码块。
for 循环可以遍历任何序列。例如，输出数组中的元素：

```
mylist = [1,2,3]
for item in mylist:
    print(item)
```

输出字符串中的字符：

```
>>> for c in 'banana' :
...     print(c,type(c))
...
b <class 'str'>
a <class 'str'>
n <class 'str'>
a <class 'str'>
n <class 'str'>
a <class 'str'>
```

因为 Python 3 中并不存在表示单个字符的数据类型，所以，返回的变量 c 仍然是 str 类型。

输出字符串 'banana' 中出现的每个字符及其位置，代码如下：

```
>>> for c in enumerate('banana'):
...     print(c)
...
(0, 'b')
(1, 'a')
(2, 'n')
(3, 'a')
(4, 'n')
(5, 'a')
```

在执行循环代码块之前，根据循环条件决定是否继续执行循环代码块，当满足循环条件时，继续执行循环体中的代码。在循环条件之前写上关键词 while。这里的 while 就是"当"的意思。例如，当用户直接输入回车时退出循环，代码如下：

```
import sys

while True:
    line = sys.stdin.readline().strip()
    if not line:
        break
    print(line)
```

2.6 列表

可以使用一个列表（list）存储任何类型的对象，示例代码如下：

```
list1 = ['physics', 'chemistry', 1997, 2000];
list2 = [1, 2, 3, 4, 5, 6, 7 ];
print("list1[0]: ", list1[0])
print("list2[1:5]: ", list2[1:5])
```

输出如下：

```
list1[0]:  physics
list2[1:5]:  [2, 3, 4, 5]
```

此外，列表还可以将另一个列表作为项目，这称为嵌套列表，示例代码如下：

```
my_list = ["mouse", [8, 4, 6], ['a']]    # 嵌套列表
```

使用 range 函数可以生成列表，示例代码如下：

```
>>> list(range(10))              # 从 0 开始，到 10, 步长为1
[0, 1, 2, 3, 4, 5, 6, 7, 8, 9]
>>> list(range(0, 30, 5))        # 从 0 开始，到 30, 步长为5
[0, 5, 10, 15, 20, 25]
```

可以使用赋值运算符（=）更改一个项目或项目范围，示例代码如下：

```
odd = [2, 4, 6, 8]               # 错误的值

odd[0] = 1                       # 改变第一项

print(odd)                       # 输出：[1, 4, 6, 8]

odd[1:4] = [3, 5, 7]             # 改变第 2 至第 4 项

print(odd)                       # 输出：[1, 3, 5, 7]
```

可以使用 append() 方法将一个项添加到列表中，或使用 extend() 方法添加多个项，示例代码如下：

```
odd = [1, 3, 5]

odd.append(7)

print(odd)                       # 输出：[1, 3, 5, 7]

odd.extend([9, 11, 13])

print(odd)                       # 输出：[1, 3, 5, 7, 9, 11, 13]
```

可以使用 + 运算符来连接两个列表，其中，* 运算符重复列表给定次数，示例代码如下：

```
odd = [1, 3, 5]

print(odd + [9, 7, 5])           # 输出：[1, 3, 5, 9, 7, 5]

print(["re"] * 3)                # 输出：["re", "re", "re"]
```

此外，还可以使用 insert() 方法在所需位置插入一个项目，或者通过将多个项目挤压到列表的空白切片中来插入多个项目，示例代码如下：

```
odd = [1, 9]
odd.insert(1,3)

print(odd)                       # 输出：[1, 3, 9]

odd[2:2] = [5, 7]
```

```
print(odd)                    # 输出: [1, 3, 5, 7, 9]
```

使用关键字 del 可以从列表中删除一个或多个项目，示例代码如下：

```
my_list = ['p','r','o','b','l','e','m']

del my_list[2]                # 删除一个项目

print(my_list)                # 输出: ['p', 'r', 'b', 'l', 'e', 'm']

del my_list[1:5]              # 删除多个项目

print(my_list)                # 输出: ['p', 'm']
```

还可以完全删除列表，示例代码如下：

```
del my_list                   # 删除整个列表

print(my_list)                # 错误: 列表未定义
```

使用 remove() 方法可以删除给定的项目。使用 pop() 方法可以删除给定索引处的项目。使用 clear() 方法可以清空列表。示例代码如下：

```
my_list = ['p','r','o','b','l','e','m']
my_list.remove('p')

print(my_list)                # 输出: ['r', 'o', 'b', 'l', 'e', 'm']

print(my_list.pop(1))         # 输出: 'o'

print(my_list)                # 输出: ['r', 'b', 'l', 'e', 'm']

print(my_list.pop())          # 输出: 'm'

print(my_list)                # 输出: ['r', 'b', 'l', 'e']

my_list.clear()

print(my_list)                # 输出: []
```

还可以通过为一个元素片段分配一个空列表来删除列表中的项目，示例代码如下：

```
>>> my_list = ['p','r','o','b','l','e','m']
>>> my_list[2:3] = []
>>> my_list
['p', 'r', 'b', 'l', 'e', 'm']
```

```
>>> my_list[2:5] = []
>>> my_list
['p', 'r', 'm']
```

for-in 语句可以轻松遍历列表中的项目，示例代码如下：

```
for fruit in ['apple','banana','mango']:
    print("I like",fruit)
```

为了复制出一个新的列表，可以使用内置的 list.copy() 方法（从 Python 3.3 开始提供），示例代码如下：

```
>>> old_list = [1, 2, 3]
>>> new_list = old_list.copy()
```

使用 new_list = my_list，实际上没有两个列表。赋值仅复制对列表的引用，而不是实际列表，因此，new_list 和 my_list 在赋值后引用相同的列表。

通常，我们只想收集符合特定条件的项目。下面有一个单词列表，我们只想从中提取包含 'wo' 的单词。为此，需要先创建一个新的空列表，然后遍历原始列表以查找要放入的项目，代码如下：

```
>>> wood = 'How much wood would a woodchuck chuck if a woodchuck could
chuck wood?'.split()
>>> wood
['How', 'much', 'wood', 'would', 'a', 'woodchuck', 'chuck', 'if', 'a',
'woodchuck', 'could', 'chuck', 'wood?']
>>> wolist = []                              # 创建一个空的列表
>>> for x in wood:
        if 'wo' in x:
            wolist.append(x)                 # 向列表增加项目
>>> wolist
['wood', 'would', 'woodchuck', 'woodchuck', 'wood?']
```

打印列表的内容，代码如下：

```
>>> mylist = ['x', 3, 'b']
>>> print('[%s]' % ', '.join(map(str, mylist)))
[x, 3, b]
```

2.7 元组

元组是一个不可变的 Python 对象序列。元组变量的赋值要在定义时就进行，定义时赋值之后就不允许修改了，示例代码如下：

```
tup1 = ('physics', 'chemistry', 1997, 2000);
tup2 = (1, 2, 3, 4, 5, 6, 7 );
print( "tup1[0]: ", tup1[0]);
print( "tup2[1:5]: ", tup2[1:5]);
```

通常将元组用于异构（不同）数据类型，将列表用于同类（相似）数据类型。

包含多个项目的文字元组可以分配给单个对象。当发生这种情况时，就好像元组中的项目已经"打包"到对象中，示例代码如下：

```
>>> t = ('foo', 'bar', 'baz', 'qux')
```

将元组中的元素分别赋给变量称为拆包，示例代码如下：

```
>>> (s1, s2, s3, s4) = ('foo', 'bar', 'baz', 'qux')
>>> s1
'foo'
>>> s2
'bar'
>>> s3
'baz'
>>> s4
'qux'
```

包装和拆包可以合并为一个语句，以进行复合分配，示例代码如下：

```
>>> (s1, s2, s3, s4) = ('foo', 'bar', 'baz', 'qux')
>>> s1
'foo'
>>> s2
'bar'
>>> s3
'baz'
>>> s4
'qux'
```

可以构建一个元组组成的数组，示例代码如下：

```
>>> pairs = [("a", 1), ("b", 2), ("c", 3)]
>>> for a, b in pairs:
...     print(a, b)
...
a 1
b 2
c 3
```

可以使用命名元组给元组中的元素起一个有意义的名字，示例代码如下：

```
import collections
```

```python
# 声明一个名为 Person 的命名元组，这个元组包含 name 和 age 两个键
Person = collections.namedtuple('Person', 'name age')

# 使用命名元组
bob = Person(name='Bob', age=30)
print('\nRepresentation:', bob)

jane = Person(name='Jane', age=29)
print('\nField by name:', jane.name)

print('\nFields by index:')
for p in [bob, jane]:
    print('{} is {} years old'.format(*p))
```

2.8 集合

可以使用运算符 in 来检查给定元素是否存在于集合中。如果集合中存在指定元素，则返回 True，否则返回 False，示例代码如下：

```
>>> s = {1,2,3,4,5}            # 创建 set 对象并将其分配给变量 s
>>> contains = 1 in s          # 判断是否包含的例子
>>> print(contains)
True
>>> contains = 6 in s
>>> print(contains)
False
```

输出字符串 'banana' 中的字符集合，代码如下：

```
>>> set(c for (i,c) in enumerate('banana'))
{'n', 'a', 'b'}
```

2.9 字典

字典是另一种可变容器模型，可存储任意类型对象。要访问字典元素，可以使用方括号和键来获取它的值，示例代码如下：

```
dict = {'Name': 'Zara', 'Age': 7, 'Class': 'First'}
print("dict['Name']: ", dict['Name'])
print("dict['Age']: ", dict['Age'])
```

可以按字典中的值排序，由于字典本质上是无序的，所以，可以把排序结果保存到有序的列表中，示例代码如下：

```
>>> x = {1: 2, 3: 4, 4: 3, 2: 1, 0: 0}
>>> sorted_by_value = sorted(x.items(), key=lambda kv: kv[1])
>>> print(sorted_by_value)
[(0, 0), (2, 1), (1, 2), (4, 3), (3, 4)]
```

OrderedDict 是一个字典子类，它会记住键/值对的顺序，示例代码如下：

```
import collections

print('普通的字典:')
d = {}
d['a'] = 'A'
d['b'] = 'B'
d['c'] = 'C'

for k, v in d.items():
    print(k, v)

print('\n有序的字典:')
d = collections.OrderedDict()
d['a'] = 'A'
d['b'] = 'B'
d['c'] = 'C'
d['a'] = 'a'

for k, v in d.items():
    print(k, v)
```

2.10 位数组

位数组（BitSet）是一串可以按位运算的位。PyRoaringBitMap（*https://github.com/Ezibenroc/PyRoaringBitMap*）是一个 C 语言库 CRoaring 的 Python 包装器。

可以使用 Pypi 安装 pyroaring，代码如下：

```
# pip3 install pyroaring
```

或者从 whl 文件安装，代码如下：

```
# pip3 install --user https://github.com/Ezibenroc/PyRoaringBitMap/releases/download/0.2.1/pyroaring-0.2.1-cp36-cp36m-linux_x86_64.whl
```

可以像使用经典的 Python 集合那样在代码中使用 BitMap，示例代码如下：

```python
from pyroaring import BitMap
bm1 = BitMap()
bm1.add(3)
bm1.add(18)
bm2 = BitMap([3, 27, 42])
print("bm1        = %s" % bm1)
print("bm2        = %s" % bm2)
print("bm1 & bm2  = %s" % (bm1&bm2))
print("bm1 | bm2  = %s" % (bm1|bm2))
```

输出如下：

```
bm1        = BitMap([3, 18])
bm2        = BitMap([3, 27, 42])
bm1 & bm2  = BitMap([3])
bm1 | bm2  = BitMap([3, 18, 27, 42])
```

遍历位数组，代码如下：

```
>>> a = iter(bm1)                   # 取得 iterator
>>> print(next(a, None))            # 取得下一个元素，如果没有则返回 None
3
>>> print(next(a, None))
18
>>> print(next(a, None))
None
```

2.11 模块

可以使用 import 语句导入一个 .py 文件中定义的函数。一个 .py 文件就称为一个模块（Module）。例如，存在一个 re.py 文件，可以使用 import re 语句导入这个正则表达式模块。

使用正则表达式模块去掉一些标点符号的示例代码如下：

```python
import re

line = 'Hi.'
normtext = re.sub(r'[\.,:;\?]', '', line)
print(normtext)
```

从 re 模块直接导入 sub 函数的示例代码如下：

```python
from re import sub
```

```
line = 'Hi.'
normtext = sub(r'[\.,:;\?]', '', line)
print(normtext)
```

模块越来越多以后，会难以管理。例如，可能会出现重名的模块。一个班里有两个名为陈晨的同学。如果他们在不同的小组，可以叫第一组的陈晨或者第三组的陈晨，这样就能区分同名了。为了避免名字冲突，模块可以位于不同的命名空间，叫作包。可以在模块名前面加上包名限定，这样即使模块名相同，也不会冲突了。

为了查看本地有哪些模块可用，可以在 Python 交互式环境中输入如下代码：

```
help('modules')
```

2.12 函数

可以把一段多次重复出现的函数命名成一个有意义的名字，然后通过名字来执行这段代码。有名字的代码段就是一个函数。使用关键字 def 可以定义一个函数，示例代码如下：

```
def square(number):           # 定义一个名为 square 的函数
    return number * number    # 返回一个数的平方
print(square(3))              # 输出：9
```

代码中可以给函数增加说明，示例代码如下：

```
def square_root(n):
    """ 计算一个数字的平方根。

    Args:
        n: 用来求平方根的数字。
    Returns:
        n 的平方根。
    Raises:
        TypeError: 如果 n 不是数字。
        ValueError: 如果 n 是负数。

    """
    pass
```

参数可以有默认值，例如，定义一个名为 RunKaldiCommand 的函数，示例代码如下：

```
import subprocess
```

```
def RunKaldiCommand(command, wait = True):      # wait 的默认值是 True
    """ 通常执行由管道连接的一系列命令，所以我们使用 shell=True """
    p = subprocess.Popen(command, shell = True,
                         stdout = subprocess.PIPE,
                         stderr = subprocess.PIPE)

    if wait:
        [stdout, stderr] = p.communicate()
        if p.returncode is not 0:               # 执行命令出现错误
            raise Exception("There was an error while running the command {0}\n".format(command)+"-"*10+"\n"+stderr)
        return stdout, stderr
    else:
        return p
```

使用如下函数：

```
RunKaldiCommand("ls -lh")
```

这里只给 RunKaldiCommand 方法的第一个参数传递了值，第二个值采用默认的 True。如果需要声明可变数量的参数，可在这个参数前面加 *，示例代码如下：

```
def myFun(*argv):
    for arg in argv:
        print (arg)

myFun('Hello', 'a', 'to', 'b')
```

函数定义中的特殊语法 **kwargs 用于传递一个键 / 值对的可变长度的参数列表，示例代码如下：

```
def myFun(**kwargs):
    for key, value in kwargs.items():
        print ("%s == %s" %(key, value))

# 调用函数
myFun(first ='test', mid ='for', last='abc')
```

输出结果如下：

```
first == test
mid == for
last == abc
```

每个 python 文件 / 脚本（模块）都有一些未明确声明的内部属性。其中一个属性是 __builtins__ 属性，它本身包含许多有用的属性和功能。可以在这里找到 __name__ 属性，根据模块的使用方式，它可以具有不同的值。

当把 python 模块作为程序直接运行时（无论是从命令行还是双击它），如果当前模块是 Python 程序的入口模块（也称顶级模块、脚本文件），则当前模块的 __name__ 属性的值是 "__main__"。

相比之下，当一个模块被导入到另一个模块中（或者在 python REPL 被导入）时，__name__ 属性中的值是模块本身的名称（即隐式声明它的 python 文件 / 脚本的名称）。

Python 脚本是自上而下的执行。指令在解释器读取它们时执行。这可能是一个问题，如果你想要做的就是导入模块（使用 import module 方法）并利用它的一个或两个方法（使用 from ... import ... 方式），可有条件地执行这些指令——将它们包装在一个 if 语句块中。

这是 'main 函数'的目的。它是一个条件块，因此，除非满足给定的条件，否则不会处理 main 函数。

main 函数的示例代码如下：

```
import sys

def main():
    if len(sys.argv) != 2:
        sys.stderr.write("Usage: {0} <min-count>\n".format(sys.argv[0]))
        raise SystemExit(1)

    words = {}
    for line in sys.stdin.readlines():
        parts = line.strip().split()
        words[parts[1]] = words.get(parts[1], 0) + int(parts[0])

    for word, count in words.iteritems():
        if count >= int(sys.argv[1]):
            print ("{0} {1}".format(count, word))

if __name__ == '__main__':
    main()
```

2.13　print 函数

显示某个目录下的文件数量的代码如下：

```
import os

folderlist = os.listdir('/home/soft/kaldi/')
total_num_file = len(folderlist)
```

```python
print ('total '+total_num_file+' files')
```

这样会出错,因为 Python 不支持 + 运算中的整数自动转换成字符串。可以调用 str() 函数将整数转换成字符串,示例代码如下:

```python
print ('total '+str(total_num_file)+' files')
```

或者格式化,代码如下:

```python
print ('total have %d files' % (total_num_file))    #%d 表示输出整数
```

另一种格式化输出的方法是使用 str.format() 方法,如下代码比较了这两种方法。

```
>>> sub1 = "python string!"
>>> sub2 = "an arg"
>>> a = "i am a %s" % sub1
>>> b = "i am a {0}".format(sub1)
>>> print(a)
i am a python string!
>>> print(b)
i am a python string!
>>> c = "with %(kwarg)s!" % {'kwarg':sub2}
>>> print(c)
with an arg!
>>> d = "with {kwarg}!".format(kwarg=sub2)
>>> print(d)
with an arg!
如下的代码会出错:
>>> name=(1, 2, 3)
>>> print("hi there %s" % name)
Traceback (most recent call last):
  File "<stdin>", line 1, in <module>
TypeError: not all arguments converted during string formatting
```

print 函数用到的格式化字符串的约定如表 2-3 所示。

表 2-3 print 函数用到的格式化字符串的约定

转换类型	含义
d,i	带符号的十进制整数
o	不带符号的八进制
u	不带符号的十进制
x	不带符号的十六进制(小写)
X	不带符号的十六进制(大写)

续表

转换类型	含 义
e	科学记数法表示的浮点数（小写）
E	科学记数法表示的浮点数（大写）
f,F	十进制浮点数
g	如果指数大于 −4 或者小于精度值，则和 e 相同，其他情况和 f 相同
G	如果指数大于 −4 或者小于精度值，则和 E 相同，其他情况和 F 相同
C	单字符（接收整数或者单字符字符串）
r	字符串（使用 repr() 转换任意 python 对象）
s	字符串（使用 str() 转换任意 python 对象）

2.14 正则表达式

re 模块是 Python 的标准库，用于处理正则表达式的所有事情。与任何其他模块一样，可以从导入 re 开始，代码如下：

```
>>> import re
```

假设想在下面这个非常简短的文本中找到以 'wo' 开头的所有单词。想要使用的是 re.findall() 方法。它需要两个参数：①正则表达式模式；②用于查找匹配的目标字符串，代码如下：

```
>>> wood = 'How much wood would a woodchuck chuck if a woodchuck could chuck wood?'
>>> re.findall(r'wo\w+', wood)           # r'...' 表示原始字符串
['wood', 'would', 'woodchuck', 'woodchuck', 'wood']
>>>
```

首先，注意正则表达式 r'wo\w+' 使用原始字符串表示，如 r'...' 字符串前缀所示。这是因为，正则表达式使用反斜杠（\）作为它们自己的特殊转义字符，没有 'r' 时反斜杠被解释为 Python 的特殊转义字符。原始字符串以大写字母 R 或者小写字母 r 开始。

findall() 方法将所有匹配的字符串部分作为列表返回。如果没有匹配项，则返回一个空列表，代码如下：

```
>>> re.findall(r'o+', wood)
['o', 'oo', 'o', 'oo', 'oo', 'o', 'oo']
>>> re.findall(r'e+', wood)
[]
```

如果想忽略匹配中的大小写该怎么办？可以将其指定为 findall() 方法的第三个可选参数：re.IGNORECASE。

```
>>> foo = 'This and that and those'
>>> re.findall(r'th\w+', foo)
['that', 'those']
>>> re.findall(r'th\w+', foo, re.IGNORECASE)    # case is ignored while matching
['This', 'that', 'those']
```

如果想用其他东西替换所有匹配的部分怎么办？可以使用 re.sub() 方法完成。下面找到所有元音序列并用 '-' 替换。该方法将结果作为新字符串返回，代码如下：

```
>>> wood
'How much wood would a woodchuck chuck if a woodchuck could chuck wood?'
>>> re.sub(r'[aeiou]+', '-', wood)          #3 个参数：正则表达式、替换字符串、目标字符串
'H-w m-ch w-d w-ld - w-dch-ck ch-ck -f - w-dch-ck c-ld ch-ck w-d?'
```

删除匹配部分也可以通过 re.sub() 实现，只需将替换字符串设为空字符串 '' 即可，代码如下：

```
>>> re.sub(r'[aeiou]+', '', wood)           #用空字符串替换
'Hw mch wd wld  wdchck chck f  wdchck cld chck wd?'
```

如果必须在许多不同的字符串中匹配正则表达式，最好将正则表达式构造为 Python 对象。这样，正则表达式的有限状态自动机被编译一次并重复使用。由于构建 FSA（Finite State Automata）在计算上相当昂贵，因此减轻了处理负荷。为此，请使用 re.compile() 方法，代码如下：

```
>>> myre = re.compile(r'\w+ou\w+')          # 将 myre 编译为正则表达式对象
>>> myre.findall(wood)                      # 直接在 myre 上调用 .findall() 方法
['would', 'could']
>>> myre.findall('Colorless green ideas sleep furiously')
['furiously']
>>> myre.findall('The thirty-three thieves thought that they thrilled the throne throughout Thursday.')
['thought', 'throughout']
```

编译完成后，就可以直接在正则表达式对象上调用一个 re 方法。在上面的例子中，myre 是对应于 r'\w+ou\w+' 的编译正则表达式对象，在 myre 上调用 .findall() 方法。相对于没编译时，现在需要指定少一个参数：目标字符串 myre.findall(wood) 是唯一需要的东西。

有时，我们只对确认给定字符串中是否存在匹配感兴趣。在这种情况下，re.search() 是一个很好的选择。re.search() 方法仅查找第一个匹配，然后退出。如果找到匹配项，则返回一个匹配对象。如果没有，则不会返回任何值。r'e+' 在 'Colorless ...' 字符串中成功

匹配，因此，返回匹配对象。Wood 字符串没有一个 'e'，所以，没有返回任何值。

```
>>> re.search(r'e+', 'Colorless green ideas sleep furiously')
<_sre.SRE_Match object at 0x02D9CB48>
>>> re.search(r'e+', wood)
>>>
```

可以在 if 条件语句的上下文中使用 re.search()。在下面的代码中，if ... 行检查 re.search 方法是否有返回的对象，然后打印匹配的部分和匹配的行（注意：if 条件判断中的 someobj 返回 True，只要 someobj 不是以下之一："nothing"、整数 0、空字符串 ""、空列表 [] 和空字典 {}）。

```
>>> f = open('D:\\Lab\\warpeace.txt')
>>> blines = f.readlines()
>>> f.close()
>>> smite = re.compile(r'sm(i|o)te\w*')
>>> for b in blines:
        matchobj = smite.search(b)
        if matchobj:            # True if matchobj is not "nothing"
            print(matchobj.group(), '-', b, end='')
```

输出如下：

```
smite - servants to rejoice in Thy mercy; smite down our enemies and destroy
smote - wives and children." The nobleman smote his breast. "We will all
smote - you, Papa" (he smote himself on the breast as a general he had heard
smite - Faith, the sling of the Russian David, shall suddenly smite his head
```

在这个例子中，我们想要找出"战争与和平"中所有含有"smite / smote ..."字样的行。首先使用 .readlines() 方法将文本文件作为行的列表加载。然后，因为多次进行匹配，所以，编译正则表达式。最后，循环遍历文本行，通过 .search() 方法创建一个匹配对象，并且只有匹配对象存在时才打印出匹配的部分和行。

2.15 文件操作

文件的绝对路径由目录和文件名两部分构成，示例代码如下：

```
import os.path

path = '/home/data/file.wav'

print(os.path.abspath(path))      # 返回绝对路径（包含文件名的全路径）
print(os.path.basename(path))     # 返回路径中包含的文件名
```

```
print(os.path.dirname(path))     # 返回路径中包含的目录
```

输出如下:

```
/home/data/file.wav
file.wav
/home/data
```

2.15.1 读写文件

逐行读入文本文件,代码如下:

```
lexicon = open("lexicon.txt")

for line in lexicon:
    line = line.strip()
    print(line,"\n")

lexicon.close()
```

读入 utf-8 编码格式的文本文件,代码如下:

```
import codecs
import sys

transcript = codecs.open(sys.argv[1], "r", "utf8")     # 第一个参数传入文件名

for line in transcript:
    print(line)

transcript.close()
```

为了实现写入文本文件,可以使用 'w' 模式的 open() 函数以写模式打开新文件,代码如下:

```
new_path = "a.speaker_info"
fout = open(new_path,'w')
```

需要注意,如果 new_days.txt 在打开文件之前已经存在,则它的旧内容将被破坏,所以在使用 'w' 模式时要小心。

一旦打开新文件,可以使用写入操作 <file>.write() 将数据放入文件中。写入操作接收单个参数,该参数必须是字符串,并将该字符串写入文件。如果想要在文件中开始新行,则必须明确提供换行符,示例代码如下:

```
fout.write("\nID:\t1212")
```

关闭文件可确保磁盘上的文件和文件变量之间的连接已完成。关闭文件还可确保其他程序能够访问它们并保证用户的数据安全。所以，一定要关闭文件。使用 <file>.close() 函数关闭所有文件，代码如下：

```
fout.close()
```

使用 open() 函数创建文件对象可以使用的模式总结如下：
- 'r' 用于读取现有文件（默认值；可以省略）。
- 'w' 用于创建用于写入的新文件。
- 'a' 用于将新内容附加到现有文件。

对于 json 格式的文件，可以导入 json 模块读取文件，代码如下：

```
import json
data = json.load(open('my_file.json', 'r'))
```

演示 json 文件的内容如下：

```
{"hello":"lietu"}
```

演示读取 json 格式的文件如下：

```
>>> import json
>>> print(json.load(open('my_file.json','r')))
{u'hello': u'lietu'}
```

2.15.2　重命名文件

Linux 操作系统中的文件名区分英文大小写，而 Window 操作系统中的文件名不区分英文大小写。

可以使用 os.rename 方法重命名文件。首先使用 touch 命令创建一个空文件，代码如下：

```
# touch ./test1
```

然后把 test1 重命名为 test2

```
import os
src= 'test1'
dst= 'test2'
os.rename(src, dst)
```

2.15.3　遍历文件

使用 os.scandir 遍历一个目录。os.scandir() 方法返回一个迭代器，示例代码如下：

```
import os

with os.scandir('/home/') as entries:
    for entry in entries:
        print(entry.name)
```

这里通过 with 语句使用上下文管理器关闭迭代器,并在迭代器耗尽后自动释放已获取的资源。

只打印出一个目录下的文件,代码如下:

```
dir_entries = os.scandir('/home/')
for entry in dir_entries:
    if entry.is_file():                    # 判断项目是否文件
        print(f'{entry.name}')
```

如果想要遍历一个目录树并处理树中的文件,则可以使用 os.walk() 方法。os.walk() 默认以自上而下的方式遍历目录,代码如下:

```
import os
for root, dirs, files in os.walk("/home/"):
    for name in files:
        print(os.path.join(root, name))    # 打印文件
    for name in dirs:
        print(os.path.join(root, name))    # 打印目录
```

2.16　with 语句

with 语句在 Python 脚本中被广泛使用。

with 语句格式如下:

```
with context [as var]:
    pass
```

这里 context 是一个表达式,它将返回一个对象并保存在 var 中。

以下是一个示例:

```
with open("data.txt") as f:
    print(type(f))
```

在本例中,open("data.txt") 将返回一个 _io.TextIOWrapper 对象,该对象将保存到变量 f 中。

使用 with 语句的主要原因是 with 语句完成后将执行一些额外的操作,示例代码如下:

```
with open("data.txt") as f:
    print(type(f))
print(f.closed)
print("--end--")
```

运行这个 Python 脚本，将得到如下结果：

```
<class '_io.TextIOWrapper'>
True
--end--
```

从上面的输出中可以发现，with 语句将在完成时关闭文件。用户不需要手动关闭此文件。还可以发现，with 语句创建的变量是全局的。在上面的例子中，变量 f 在整个 Python 脚本中都能很好地工作，而不仅仅是在 with 语句中。

2.17 使用 pickle 模块序列化对象

可以把内存中的 Python 数据对象保存成二进制文件，这样下次需要它时就可以简单地加载它并获得原始对象。该过程称为"对象序列化"。

从文件恢复出来对象称为反序列化。切勿反序列化从不受信任的来源收到的数据，因为这可能会带来一些严重的安全风险。pickle 模块在挑选恶意数据时无法知道或引发错误。

如下是一个序列化字典的简单例子。

```
import pickle
emp = {1:"A",2:"B",3:"C",4:"D",5:"E"}
pickling_on = open("Emp.pickle","wb")           #打开文件"Emp.pickle"
pickle.dump(emp, pickling_on)
pickling_on.close()                             #关闭文件
```

注意使用"wb"而不是"w"，因为所有操作都是使用字节完成的。

下面研究一下如何反序列化这个字典。

```
pickle_off = open("Emp.pickle","rb")    #读取字节时，请注意"rb"而不是"r"的用法
emp = pickle.load(pickle_off)
print(emp)
```

2.18 面向对象编程

语音识别软件往往由很多代码组成，也是一个复杂的系统，为了能够封装细节，需要

抽象出对象。对象只是数据（变量）和作用于这些数据的方法（函数）的集合。类本质上是用于创建对象的模板。

就好像函数定义以关键字 def 开头一样，在 Python 中，使用关键字 class 定义一个类。如下是一个简单的类定义。

```
class MyNewClass:
    '''This is a docstring. I have created a new class'''
    pass
```

一个类创建一个新的本地命名空间，其中定义了所有属性。属性可以是数据或函数。其中还有一些特殊属性，这些属性以双下画线（__）开头。例如，__doc__ 提供了该类的文档字符串，代码如下：

```
class MyClass:
    "This is my second class"
    a = 10
    def func(self):
        print('Hello')

print(MyClass.a)            # 输出: 10
print(MyClass.func)         # 输出: <function MyClass.func at 0x0000000003079BF8>
print(MyClass.__doc__)      # 输出: This is my second class
```

可以根据类模板来创建对象，创建对象的过程类似于函数调用，代码如下：

```
ob = MyClass()
```

这将创建一个名为 ob 的新实例对象。可以使用对象名称前缀来访问对象的属性，代码如下：

```
ob.func()  # 输出: Hello
```

读者可能已经注意到类中函数定义中的 self 参数，将该方法简单地称为 ob.func() 而没有任何参数，它仍然奏效。

这是因为，只要对象调用其方法，对象本身就作为第一个参数传递。因此，ob.func() 将转换为 MyClass.func(ob)。

方法与对象实例或类相关联，函数则不是。当 Python 调度（调用）一个方法时，它会将该调用的第一个参数绑定到相应的对象引用（对于大多数方法，这个参数通常称为 self）。

在 Python 中，除了用户定义的属性，每个对象都有一些默认属性和方法。要查看对象的所有属性和方法，可以使用内置的 dir() 函数，示例代码如下：

```
ob = MyClass()

print(dir(ob))
```

2.19 命令行参数

在采用多种编程语言开发的语音识别系统中，Python 脚本可能需要从命令行直接读取参数。如果脚本很简单或临时使用，没有多个复杂的参数选项，可以直接用 sys.argv 读取传入的命令行参数。

测试代码 TestArgv.py 内容如下：

```
import sys

print("This is the path of the script: ", sys.argv[0])   #脚本的相对路径
print("Number of arguments: ", len(sys.argv))            #长度最少是 1
print("The arguments are: " , str(sys.argv))             #str 函数输出 sys.argv 的内容
```

输出结果如下：

```
D:\PycharmProjects\Scripts\python.exe D:/PycharmProjects/untitled/TestArgv.py a b c
This is the path of the script:  D:/PycharmProjects/untitled/TestArgv.py
Number of arguments:  4
The arguments are:  ['D:/PycharmProjects/untitled/TestArgv.py', 'a', 'b', 'c']
```

相关的规范有 GNU getopt_long()。getopt_long() 是 GNU 版本的 getopt 模块中的一个函数，用于解析命令行选项和参数。在 Python 中，通常使用 argparse 模块来实现相同的功能，因为 getopt 模块并不是所有的 Python 实现都提供，而 argparse 是 Python 的标准库模块。

GNU 扩展 getopt_long() 可以解析可读的多字符选项，该选项前缀为双短线，而非单个短线。双短线选项（如 --inum）可以和单个短线选项区分开 (-abc)。GNU 扩展允许带参选项有不同的形式：--name=arg。

argparse 包使得这一工作变得简单而规范。它支持 GNU getopt_long()。

您首先导入 argparse 库并初始化 ArgumentParser 的对象。此对象将保存读取命令行参数所需的所有信息。

以下是入门的基本语法：

```
import argparse

# 初始化 ArgumentParser 对象
parser = argparse.ArgumentParser()
```

可以使用 add_argument 方法添加参数。初始化 argparse.ArgumentParser() 时，可以传递几个可选参数来自定义其行为。示例代码如下：

```
import argparse

parser = argparse.ArgumentParser(description='Example application.')
```

```
parser.add_argument('integer', type=int, help='An integer.')
parser.add_argument('-f', '--foo', default='bar', choices=['bar', 'baz'],
help='Foo option.')
parser.add_argument('--flag', action='store_true', help='A boolean flag.')

args = parser.parse_args()
```

无参数运行 Testargparse.py 的输出结果如下:

```
usage: Testargparse.py [-h] [-f {bar,baz}] [--flag] integer
Testargparse.py: error: the following arguments are required: integer
```

2.20 数据库

SQLite3 是一个非常易于使用的数据库引擎。SQLite3 是独立的、无服务器的、零配置和事务性的。它非常快速且轻量级,整个数据库存储在一个磁盘文件中,可以在语音识别应用中用作内部数据存储。Python 标准库包含一个名为"sqlite3"的模块,用于处理此数据库。

使用 sqlite3 模块在内存中创建一个 SQLite 数据库,代码如下:

```
import sqlite3

conn = sqlite3.connect('example.db')      # 连接数据库
```

接下来,通过游标创建表,代码如下:

```
c = conn.cursor()
c.execute('''CREATE TABLE results (dataset text, wer float)''')
c.execute('INSERT INTO results(dataset, wer) VALUES(?, ?)', ("LibriSpeech", 0.0583))
```

为了实际上保存更改,需要调用连接对象的 .commit() 方法,代码如下:

```
conn.commit()
```

从前面创建的 results 表中获取并显示记录,代码如下:

```
# 获得所有结果
c.execute("SELECT dataset, wer FROM results ")
d = c.fetchall()

for row in d:
    print ("dataset: {}".format( row[0] ))
    print ("wer: {}".format( row[1] ))
```

可以使用 DB Browser for SQLite（https://github.com/sqlitebrowser/sqlitebrowser）查看数据库文件中的数据。

2.21 JSON 格式

JSON（JavaScript Object Notation）是一种轻量级的数据交换格式。人类很容易阅读和编写 JSON。机器也很容易解析和生成 JSON。可以用 JSON 传输由名称/值对和数组数据类型组成的数据对象。

JSON 的基本数据类型有如下 6 种。
- 数字：有符号的十进制数字，可能包含小数部分，可能使用指数 E 表示法，但不能包括非数字，如 NaN。该格式不区分整数和浮点数。
- 字符串：零个或多个 Unicode 字符的序列。字符串用双引号分隔，并支持反斜杠转义语法。
- 布尔值：为 True 或 False 的任一值。
- 数组：零个或多个值的有序列表，每个值可以是任何类型。数组使用方括号符号，元素以半角逗号分隔。
- 对象：名称/值对的无序集合，其中名称（也称为键）是字符串。由于对象旨在表示关联数组，推荐每个键在对象内是唯一的。对象用大括号分隔，并使用逗号分隔每对，而在每一对中，冒号 ':' 字符将键或名称与其值分隔开。
- null：一个空值，使用单词 null。

json.dumps() 函数将字典转换为字符串对象。例如，如下代码将输出环境变量。

```
import json, os
print(json.dumps(dict(os.environ), indent = 2))
```

json.loads() 函数将 JSON 格式的字符串解析为 Python 中的字典或列表，示例代码如下：

```
import json

r = {'is_claimed': 'True', 'rating': 3.5}
r = json.dumps(r)
loaded_r = json.loads(r)
loaded_r['rating']         # 输出 3.5
type(r)                    # 输出 str
type(loaded_r)             # 输出 dict
```

2.22 日志记录

机器学习的训练过程可能很长,为了监控运行状态,可以用日志记录,代码如下:

```
import logging

logging.basicConfig(level=logging.DEBUG)
logging.debug('trainning...')
```

日志级别大小关系如下:CRITICAL > ERROR > WARNING > INFO > DEBUG > NOTSET,当然也可以自己定义日志级别。

处理器将日志记录发送到任何输出。这些输出用自己的方式处理日志记录。

例如,FileHandler 将获取日志记录并将其附加到文件中。

标准日志记录模块已经配备了多个内置处理器,例如:

- 可以写入文件的多个文件处理器(TimeRotated、SizeRotated、Watched)。
- StreamHandler 可以输出到 stdout 或 stderr 等流。
- SMTPHandler 通过电子邮件发送日志记录。
- SocketHandler 将日志记录发送到流套接字。

此外,还有 SyslogHandler、NTEventHandler、HTTPHandler、MemoryHandler 等处理器。

格式器负责将元数据丰富的日志记录序列化为一个字符串。如果没有提供,则有一个默认格式器。记录库提供的通用格式器类将模板和样式作为输入。然后可以为日志记录对象中的所有属性声明占位符。

例如,'%(asctime)s %(levelname)s %(name)s: %(message)s' 会生成如下日志:

```
2017-07-19 15:31:13,942 INFO parent.child: Hello EuroPython。
```

注意,属性消息是使用提供的参数对日志的原始模板进行插值的结果。例如,对于 logger.info("Hello %s", "Laszlo"),消息将是 "Hello Laszlo"。

TestStreamHandler.py 中的例子代码如下:

```
import logging

logger = logging.getLogger(__name__)
logger.setLevel(logging.INFO)
handler = logging.StreamHandler()
handler.setLevel(logging.INFO)
formatter = logging.Formatter('%(asctime)s [%(filename)s:%(lineno)s - %(funcName)s - %(levelname)s ] %(message)s')
handler.setFormatter(formatter)
logger.addHandler(handler)
```

```
string = ''
logger.info("trainning... \n {0}".format(string))
```

输出结果如下:

```
2018-06-24 22:01:28,465 [TestStreamHandler.py:12 - <module> - INFO ] trainning...
```

2.23 异常处理

当人处于危险的环境中时,血液中的肾上腺素会升高。可以在运行时可能发生问题的代码中检查是否有异常发生,因为代码包装在 try 关键词中,所以称为 try 代码块。在 try 代码块中捕捉异常,而在 except 代码块中处理异常。这样把异常处理代码和正常的流程分开,就能使正常的处理流程代码能够连贯在一起。

except 代码块又称异常处理器,其常见格式如下:

```
except ExceptionClass as e:     // 异常类型
        // 处理代码
```

下列代码将捕捉路径创建中的异常:

```
import errno

output_dir = "d:/test"

try:
        os.makedirs(output_dir)
except OSError as e:
        if e.errno == errno.EEXIST and os.path.isdir(output_dir):
            print(" 路径已经存在 ");
            pass
        else:
            raise e
```

2.24 本章小结

Python 于 20 世纪 80 年代后期由荷兰的 Guido van Rossum 构思,作为 ABC 语言的继承者,能够处理异常并与阿米巴操作系统连接。Python 2 于 2000 年 10 月 16 日发布,具有许多新功能,包括循环检测垃圾收集器和对 Unicode 的支持。Python 3 于 2008 年 12 月 3 日发布。它是该语言的一个重要修订,并非完全向后兼容。它的许多主要功能都被反向移植到 Python 2.6.x 和 2.7.x 版本系列。Python 3 的发布包括 2to3 实用程序,它可以自动

（至少部分地）将 Python 2 代码转换为 Python 3。

Python 是一种多范式编程语言，完全支持面向对象的编程和结构化编程，其许多功能支持函数编程和面向切面编程。

Python 的名字源于英国喜剧组织 Monty Python（巨蟒）。Monty Python 引用经常出现在 Python 代码和文件中。例如，Python 中经常使用的伪变量是 spam 和 eggs，而不是传统的 foo 和 bar。

第 3 章 PyTorch 中的深度学习

一般而言，人工神经网络具有生物学动机，意味着它们试图模仿真实神经系统的行为。就像真正神经系统中最小的构建单元是神经元一样，人工神经网络也是如此——最小的构建单元是人工神经元。

目前，往往把神经元按层次组织成包括输入层和输出层在内的多层结构。除了输入层和输出层，神经网络还有中间层。中间层又称隐藏层或编码器。

浅层网络隐藏层的数量较少。虽然有研究表明浅层网络也可以拟合任何函数，但拟合有些函数需要非常的"宽大"，可能一层就要成千上万个神经元，这直接导致的后果是参数的数量增加到很多。深层网络能够以比浅层网络更少的参数来更好地拟合函数。

3.1 神经网络基础

本节以使用神经网络实现 XOR 运算——一个简单的线性不可分问题为例，介绍神经网络的基础知识。

3.1.1 实现深度前馈网络

深度前馈网络（Deep Feedforward Network，DFN）又称多层感知器（Multilayer Perceptron，MLP），是一种前馈人工神经网络模型，可以解决任何线性不可分问题。

实现 XOR 运算的多层感知器网络结构如下。

第一层，即输入层，有 2 个神经元。
第二层，即隐藏层，有 4 个神经元。
第三层，即输出层，有 1 个神经元。
激活函数可以选择 tanh 函数或者 sigmoid 函数等。tanh 函数的取值范围是 [-1,1]，而 sigmoid 函数的取值范围是 [0,1]。

实现代码如下：

```
import torch
import torch.nn as nn
```

```python
import numpy as np

# 定义输入和标签
X = torch.Tensor([[0, 0], [0, 1], [1, 0], [1, 1]])
y = torch.Tensor([[0], [1], [1], [0]])

# 定义神经网络模型
class XORNet(nn.Module):
    def __init__(self):
        super(XORNet, self).__init__()
        self.linear1 = nn.Linear(2, 4)
        self.linear2 = nn.Linear(4, 1)
        self.sigmoid = nn.Sigmoid()

    def forward(self, x):
        out = self.linear1(x)
        out = self.sigmoid(out)
        out = self.linear2(out)
        out = self.sigmoid(out)
        return out


net = XORNet()

# 定义损失函数和优化器
criterion = nn.MSELoss()
optimizer = torch.optim.SGD(net.parameters(), lr=0.1)

# 训练网络
for epoch in range(10000):
    optimizer.zero_grad()
    outputs = net(X)
    loss = criterion(outputs, y)
    loss.backward()
    optimizer.step()

    if (epoch + 1) % 1000 == 0:
        print(f"Epoch [{epoch + 1}/{10000}], Loss: {loss.item():.6f}")

# 预测新数据
x_test = torch.Tensor([[0, 1], [1, 0], [0, 0], [1, 1]])
outputs = net(x_test)
print(f"outputs: {outputs}")
predictions = np.around(outputs.data.numpy())
print(f"Predictions: {predictions}")
```

在上面的代码中，首先定义了输入 X 和标签 y。其中，X 是一个大小为 4×2 的张量，y 是大小为 4×1 的张量。接着定义了一个简单的神经网络模型，包含 1 个输入层、1 个包含 4 个节点的隐藏层和 1 个输出层。隐藏层使用 sigmoid 激活函数，输出层也使用 sigmoid 激活函数。定义损失函数为均方误差（MSE），优化器为随机梯度下降（SGD）算法。

在训练过程中，张量 X 和 y 被输入到模型中进行前向传播，计算模型输出和真实标签之间的均方误差，反向传播误差，并通过优化器更新模型参数。进行 10000 个 epoch 的训练后，使用模型预测新的数据。需要注意的是，由于 sigmoid 在输出层中使用，模型的输出将在 0～1，因此，需要使用 np.around() 将输出四舍五入到 0 或 1 来预测标签。

3.1.2 计算过程

前馈神经网络的基本结构如图 3-1 所示。

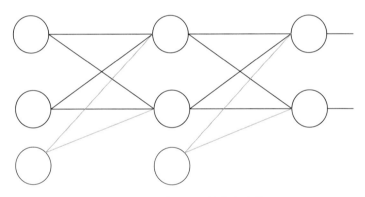

图 3-1　前馈神经网络的基本结构

使用一些数字计算，图 3-2 所示为初始的权重、偏置和训练输入 / 输出。

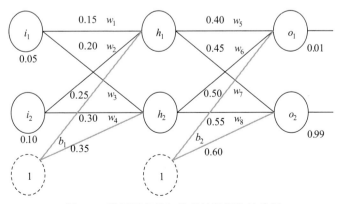

图 3-2　设置了具体权值的神经网络结构图

反向传播的目标是优化权重，以便神经网络可以学习如何正确映射任意输入到输出。本节的其余部分，将使用单个训练集：给定输入 0.05 和 0.10，希望神经网络输出 0.01 和 0.99。

首先，让我们看看神经网络在给定权重和偏置时，对于输入 0.05 和 0.10，预测的是什么。为此，将通过网络向前馈送这些输入。

计算出每个隐藏层神经元的总净输入，使用激活函数压缩总净输入（这里使用逻辑函数），然后用输出层神经元重复该过程。

计算 h_1 的总净输入：

$net_{h_1} = w_1 * i_1 + w_2 * i_2 + b_1 * 1$

$net_{h_1} = 0.15 * 0.05 + 0.2 * 0.1 + 0.35 * 1 = 0.3775$

然后使用逻辑函数对其进行压缩，以获得 h_1 的输出：

$out_{h_1} = \dfrac{1}{1+e^{-net_{h_1}}} = \dfrac{1}{1+e^{-0.3775}} = 0.593269992$

对 h_2 执行相同的过程，得

$out_{h_2} = 0.596884378$

为输出层神经元重复这个过程，使用隐藏层神经元的输出作为输入。

以下是 o_1 的输出：

$net_{o_1} = w_5 * out_{h_1} + w_6 * out_{h_2} + b_2 * 1$

$net_{o_1} = 0.4 * 0.593269992 + 0.45 * 0.596884378 + 0.6 * 1 = 1.105905967$

$out_{o_1} = \dfrac{1}{1+e^{-net_{o_1}}} = \dfrac{1}{1+e^{-1.105905967}} = 0.75136507$

对 o_2 执行相同的过程，得

$out_{o_2} = 0.772928465$

接下来计算总误差。

$E_{total} = \sum \dfrac{1}{2}(target - output)^2$

例如，o_1 的目标输出为 0.01，但神经网络输出为 0.75136507，因此其错误为

$E_{o_1} = \dfrac{1}{2}(target_{o_1} - out_{o_1})^2 = \dfrac{1}{2}(0.01 - 0.75136507)^2 = 0.274811083$

对 o_2 重复这个过程（记住目标是 0.99），得

$E_{o_2} = 0.023560026$

神经网络的总误差是这些误差的总和：

$E_{total} = E_{o_1} + E_{o_2} = 0.274811083 + 0.023560026 = 0.298371109$

使用反向传播的目标是更新网络中的每个权重，以便让实际输出更接近目标输出，从而最大限度地减少每个输出神经元和整个网络的误差。

我们想知道 w_5 的变化对总误差的影响有多大,也就是计算 $\dfrac{\partial E_{\text{total}}}{\partial w_5}$。

$\dfrac{\partial E_{\text{total}}}{\partial w_5}$ 读作"关于 w_5 的 E_{total} 的偏导数",也可以说"关于 w_5 的梯度"。

通过应用求复合函数的导数的链式法则,可以知道:

$$\frac{\partial E_{\text{total}}}{\partial w_5} = \frac{\partial E_{\text{total}}}{\partial \text{out}_{o_1}} * \frac{\partial \text{out}_{o_1}}{\partial \text{net}_{o_1}} * \frac{\partial \text{net}_{o_1}}{\partial w_5}$$

用可视化的方式展现我们正在做的事情,如图 3-3 所示。

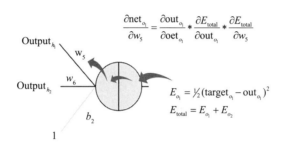

图 3-3　E_{total} 对 w_5 求偏导数

我们需要找出这个方程中的每一部分。

首先,总误差相对于输出的变化有多大?

$$E_{\text{total}} = \frac{1}{2}(\text{target}_{o_1} - \text{out}_{o_1})^2 + \frac{1}{2}(\text{target}_{o_2} - \text{out}_{o_2})^2$$

$$\frac{\partial E_{\text{total}}}{\partial \text{out}_{o_1}} = 2 * \frac{1}{2}(\text{target}_{o_1} - \text{out}_{o_1})^{2-1} * (-1) + 0$$

$$\frac{\partial E_{\text{total}}}{\partial \text{out}_{o_1}} = -(\text{target}_{o_1} - \text{out}_{o_1}) = -(0.01 - 0.75136507) = 0.74136507$$

当对 out_{o_1} 取总误差的偏导数时,数量 $\dfrac{1}{2}(\text{target}_{o_2} - \text{out}_{o_2})^2$ 变为零,因为 out_{o_1} 不会影响它,这意味着我们正在取一个常数为零的导数。

接下来,o_1 的输出相对于其总净投入量有多少变化?

逻辑函数的偏导数是输出乘以 1 减去输出:

$$\text{out}_{o_1} = \frac{1}{1 + e^{-\text{net}_{o_1}}}$$

$$\frac{\partial \text{out}_{o_1}}{\partial \text{net}_{o_1}} = \text{out}_{o_1}(1 - \text{out}_{o_1}) = 0.75136507 \times (1 - 0.75136507) = 0.186815602$$

最后,o_1 的总净输入相对于 w_5 变化多少?

$$\text{net}_{o_1} = w_5 * \text{out}_{h_1} + w_6 * \text{out}_{h_2} + b_2 * 1$$

$$\frac{\partial \text{net}_{o_1}}{\partial w_5} = 1 * \text{out}_{h_1} * w_5^{(1-1)} + 0 + 0 = \text{out}_{h_1} = 0.593269992$$

把它放在一起：

$$\frac{\partial E_{\text{total}}}{\partial w_5} = \frac{\partial E_{\text{total}}}{\partial \text{out}_{o_1}} * \frac{\partial \text{out}_{o_1}}{\partial \text{net}_{o_1}} * \frac{\partial \text{net}_{o_1}}{\partial w_5}$$

$$\frac{\partial E_{\text{total}}}{\partial w_5} = 0.74136507 * 0.186815602 * 0.593269992 = 0.082167041$$

为了减少误差，从当前权重中减去这个值（可选地乘以某个学习率 eta，合理的学习率可以使优化器快速收敛。一般在训练初期给予较大的学习率，随着训练的进行，学习率逐渐减小。这里将其设置为0.5）：

$$w_5^+ = w_5 - \eta * \frac{\partial E_{\text{total}}}{\partial w_5} = 0.4 - 0.5 * 0.082167041 = 0.35891648$$

可以重复这个过程来获得新的权重 w_6、w_7 和 w_8：

$w_6^+ = 0.408666186$

$w_7^+ = 0.511301270$

$w_8^+ = 0.561370121$

将新权重引入隐含层神经元之后，才执行神经网络中的实际更新（即当我们继续使用下面的反向传播算法时，将使用原始权重，而不是更新后的权重）。

接下来，将通过计算 w_1, w_2, w_3 和 w_4 的新值来继续向后传递。

$$\frac{\partial E_{\text{total}}}{\partial w_1} = \frac{\partial E_{\text{total}}}{\partial \text{out}_{h_1}} * \frac{\partial \text{out}_{h_1}}{\partial \text{net}_{h_1}} * \frac{\partial \text{net}_{h_1}}{\partial w_1}$$

E_{total} 对 w_1 求偏导数如图 3-4 所示。

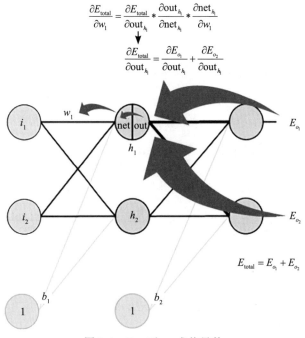

图 3-4　E_{total} 对 w_1 求偏导数

将使用与处理输出层类似的过程，但略有不同，以说明每个隐藏层神经元的输出对多个输出神经元的输出的贡献（并因此产生误差）。我们知道，out_{h_1} 同时影响 out_{o_1} 和 out_{o_2}，因此，$\dfrac{\partial E_{\text{total}}}{\partial \text{out}_{h_1}}$ 需要考虑它对两个输出神经元的影响：

$$\frac{\partial E_{\text{total}}}{\partial \text{out}_{h_1}} = \frac{\partial E_{o_1}}{\partial \text{out}_{h_1}} + \frac{\partial E_{o_2}}{\partial \text{out}_{h_1}}$$

从 $\dfrac{\partial E_{o_1}}{\partial \text{out}_{h_1}}$ 开始：

$$\frac{\partial E_{o_1}}{\partial \text{out}_{h_1}} = \frac{\partial E_{o_1}}{\partial \text{net}_{o_1}} * \frac{\partial \text{net}_{o_1}}{\partial \text{out}_{h_1}}$$

可以使用之前计算的值来计算 $\dfrac{\partial E_{o_1}}{\partial \text{net}_{o_1}}$：

$$\frac{\partial E_{o_1}}{\partial \text{net}_{o_1}} = \frac{\partial E_{o_1}}{\partial \text{out}_{o_1}} * \frac{\partial \text{out}_{o_1}}{\partial \text{net}_{o_1}} = 0.74136507 * 0.186815602 = 0.138498562 \text{，并且} \frac{\partial \text{net}_{o_1}}{\partial \text{out}_{h_1}} \text{等于} w_5\text{，}$$

推导过程如下：

$$\text{net}_{o_1} = w_5 * \text{out}_{h_1} + w_6 * \text{out}_{h_2} + b_2 * 1$$

$$\frac{\partial \text{net}_{o_1}}{\partial \text{out}_{h_1}} = w_5 = 0.40$$

代入这些值：

$$\frac{\partial E_{o_1}}{\partial \text{out}_{h_1}} = \frac{\partial E_{o_1}}{\partial \text{net}_{o_1}} * \frac{\partial \text{net}_{o_1}}{\partial \text{out}_{h_1}} = 0.138498562 * 0.40 = 0.055399425$$

对于 $\dfrac{\partial E_{o_2}}{\partial \text{out}_{h_1}}$，按照相同的过程，得

$$\frac{\partial E_{o_2}}{\partial \text{out}_{h_1}} = -0.019049119$$

因此

$$\frac{\partial E_{\text{total}}}{\partial \text{out}_{h_1}} = \frac{\partial E_{o_1}}{\partial \text{out}_{h_1}} + \frac{\partial E_{o_2}}{\partial \text{out}_{h_1}} = 0.055399425 + -0.019049119 = 0.036350306 \text{现在我们有} \frac{\partial E_{\text{total}}}{\partial \text{out}_{h_1}}\text{，}$$

我们需要计算出 $\dfrac{\partial \text{out}_{h_1}}{\partial \text{net}_{h_1}}$，然后为每个权重计算出 $\dfrac{\partial \text{net}_{h_1}}{\partial w}$：

$$\text{out}_{h_1} = \frac{1}{1+\text{e}^{-\text{net}_{h_1}}}$$

$$\frac{\partial \text{out}_{h_1}}{\partial \text{net}_{h_1}} = \text{out}_{h_1}(1-\text{out}_{h_1}) = 0.59326999 \times (1-0.59326999) = 0.241300709$$

计算总的网络输入到 h_1 对于 w_1 的偏导数：

$$\text{net}_{h_1} = w_1 * i_1 + w_3 * i_2 + b_1 * 1$$

$$\frac{\partial \text{net}_{h_1}}{\partial w_1} = i_1 = 0.05$$

把这些值放在一起：

$$\frac{\partial E_{\text{total}}}{\partial w_1} = \frac{\partial E_{\text{total}}}{\partial \text{out}_{h_1}} * \frac{\partial \text{out}_{h_1}}{\partial \text{net}_{h_1}} * \frac{\partial \text{net}_{h_1}}{\partial w_1}$$

$$\frac{\partial E_{\text{total}}}{\partial w_1} = 0.036350306 * 0.241300709 * 0.05 = 0.000438568$$

现在可以更新 w_1：

$$w_1^+ = w_1 - \eta * \frac{\partial E_{\text{total}}}{\partial w_1} = 0.15 - 0.5 * 0.000438568 = 0.149780716$$

对 w_2，w_3 和 w_4 重复此操作：

$w_2^+ = 0.19956143$

$w_3^+ = 0.24975114$

$w_4^+ = 0.29950229$

最后，已经更新了所有的权重。当最初输入 0.05 和 0.1 时，网络上的误差为 0.298371109。在第一轮反向传播之后，总误差降至 0.291027924。下降的可能看起来并不多，但是在重复此过程 10 000 次之后，例如，错误直线下降到 0.0000351085。此时，当输入 0.05 和 0.1 时，两个输出神经元产生 0.015912196（和 0.01 的目标值相比较）和 0.984065734（和 0.99 的目标值相比较）。

3.2 卷积神经网络

随着深度神经网络技术的成熟和发展，往往采用层数很深的神经网络来识别图像。

输入层和隐层之间是通过权值连接起来的，如果把输入层和隐层的神经元全部连接起来，那么权值数量就太多了。例如，对于一幅 1 000 像素 ×1 000 像素大小的图像，输入层的神经元的个数就是每个像素点，个数为 1 000×1 000 个。假设和这个输入层连接的隐层的神经元个数为 1 000 000 个，那么 W 的数量就是 1 000×1 000×1 000 000。

对于给定的输入图片，用一个卷积核处理这张图片，也就是说一个卷积核处理整张图，所以权重是一样的，这称为权值共享。

卷积层提取出特征，再进行组合形成更抽象的特征，最后形成对图片对象的描述特征。

卷积神经网络（Convolutional Neural Network，CNN）具有独特的结构，旨在模仿真实动物大脑的运作方式，而不是让每层中的每个神经元连接到下一层中的所有神经元（多层感知器），神经元以三维结构排列，以便考虑不同神经元之间的空间关系。

我们将从探索什么是卷积神经网络及其工作方式开始。然后，将使用 torchvision（一个包含各种数据集和与计算机视觉相关的辅助函数的库）加载 CIFAR10 数据集。然后，从头开始构建和训练 CNN。最后，将测试我们的模型。

卷积神经网络（CNN）获取输入图像并将其分类为任何输出类。每个图像都经过一系列不同的层——主要是卷积层、池化层和完全连接层。

卷积层用于从输入图像中提取特征。它是输入图像和内核（滤波器）之间的数学运算。使用不同的滤波器来提取不同种类的特征。

池化层用于减小任何图像的大小，同时保持最重要的特征。使用的最常见类型的池化层是最大池化和平均池化，它们分别从给定大小的滤波器（即 2×2、3×3 等）中获得最大值和平均值。

从加载一些数据开始。我们将使用 CIFAR-10 数据集。该数据集有 60 000 个 32px × 32px 的彩色图像，属于 10 个不同的类别（6 000 个图像 / 类别）。数据集分为 50 000 个训练图像和 10 000 个测试图像。

从导入所需的库并定义一些变量开始，代码如下：

```
# Load in relevant libraries, and alias where appropriate
import torch
import torch.nn as nn
import torchvision
import torchvision.transforms as transforms

# Define relevant variables for the ML task
batch_size = 64
num_classes = 10
learning_rate = 0.001
num_epochs = 20

# Device will determine whether to run the training on GPU or CPU.
device = torch.device('cuda' if torch.cuda.is_available() else 'cpu')
```

device 变量将决定是在 GPU 上还是在 CPU 上运行训练。

为了加载数据集，将使用 torchvision 中的内置数据集。它提供了下载数据集的能力，还可以应用想要的任何转换。

先来看代码：

```
# Use transforms.compose method to reformat images for modeling,
# and save to variable all_transforms for later use
all_transforms = transforms.Compose([transforms.Resize((32,32)),
                                     transforms.ToTensor(),
                                     transforms.Normalize(mean=[0.4914, 0.4822, 0.4465],
                                                          std=[0.2023, 0.1994, 0.2010])
                                     ])
# Create Training dataset
train_dataset = torchvision.datasets.CIFAR10(root = './data',
```

```
                                        train = True,
                                        transform = all_transforms,
                                        download = True)

# Create Testing dataset
test_dataset = torchvision.datasets.CIFAR10(root = './data',
                                        train = False,
                                        transform = all_transforms,
                                        download=True)

# Instantiate loader objects to facilitate processing
train_loader = torch.utils.data.DataLoader(dataset = train_dataset,
                                        batch_size = batch_size,
                                        shuffle = True)

test_loader = torch.utils.data.DataLoader(dataset = test_dataset,
                                        batch_size = batch_size,
                                        shuffle = True)
```

剖析一下这段代码：

①从写一些转换开始。调整图像的大小，将其转换为张量，并通过使用输入图像中每个频带的平均值和标准差对其进行归一化。

②加载数据集，包括训练和测试。将 download 参数设置为 True，以便在尚未下载的情况下进行下载。

③一次将整个数据集加载到内存中不是一个好的做法，严重的话，可能会导致计算机停机。这就是使用数据加载器的原因，它允许通过批量加载数据来迭代数据集。

④创建两个数据加载程序（用于训练/测试），并设置批次大小，将 shuffle 设置为 True，以便每个类的图像都包含在一个批次中。

在深入研究代码之前，了解一下如何在 PyTorch 中定义神经网络。

①创建一个新类，该类扩展了 PyTorch 中的 nn.Module 类。当创建神经网络时，这是必要的，因为它提供了一系列有用的方法。

②必须定义神经网络中的层，这是在类的 __init__ 方法中完成的。只是简单地命名层，然后将它们分配给想要的适当层，如卷积层、池化层、全连接层等。

③最后要做的事情是在类中定义一个 forward() 方法。此方法的目的是定义输入数据通过各个层的顺序。

现在，深入了解代码。

```
# Creating a CNN class
class ConvNeuralNet(nn.Module):
    # Determine what layers and their order in CNN object
```

```python
    def __init__(self, num_classes):
        super(ConvNeuralNet, self).__init__()
        self.conv_layer1 = nn.Conv2d(in_channels=3, out_channels=32, kernel_size=3)
        self.conv_layer2 = nn.Conv2d(in_channels=32, out_channels=32, kernel_size=3)
        self.max_pool1 = nn.MaxPool2d(kernel_size = 2, stride = 2)

        self.conv_layer3 = nn.Conv2d(in_channels=32, out_channels=64, kernel_size=3)
        self.conv_layer4 = nn.Conv2d(in_channels=64, out_channels=64, kernel_size=3)
        self.max_pool2 = nn.MaxPool2d(kernel_size = 2, stride = 2)

        self.fc1 = nn.Linear(1600, 128)
        self.relu1 = nn.ReLU()
        self.fc2 = nn.Linear(128, num_classes)

    # Progresses data across layers
    def forward(self, x):
        out = self.conv_layer1(x)
        out = self.conv_layer2(out)
        out = self.max_pool1(out)

        out = self.conv_layer3(out)
        out = self.conv_layer4(out)
        out = self.max_pool2(out)

        out = out.reshape(out.size(0), -1)

        out = self.fc1(out)
        out = self.relu1(out)
        out = self.fc2(out)
        return out
```

首先创建一个继承 nn.Module 类的类，然后分别在 __init__() 和 forward() 中定义层及其执行顺序。

这里需要注意如下事项：

- nn.Conv2d() 用于定义卷积层。这里定义了它们接收的通道，以及它们应该返回多少通道及内核大小。这里从 3 个通道开始，因为使用 RGB 图像。
- nn.MaxPool2d() 是一个最大池化层，只需要内核大小和步长。
- nn.Lineral() 是完全连接的层，nn.ReLU() 是使用的激活函数。
- 在 forward() 方法中定义序列，在完全连接层之前，重塑输出，使输入与完全连接层匹配。

现在为训练目的设置一些超参数，代码如下：

```
model = ConvNeuralNet(num_classes)
```

```python
# Set Loss function with criterion
criterion = nn.CrossEntropyLoss()

# Set optimizer with optimizer
optimizer = torch.optim.SGD(model.parameters(), lr=learning_rate, weight_decay = 0.005, momentum = 0.9)

total_step = len(train_loader)
```

首先，用类的数量初始化模型；然后，分别选择交叉熵和 SGD（随机梯度下降）作为损失函数和优化器。这些有不同的选择，但在实验时可以获得最大的准确性。这里还定义了变量 total_step，以使通过各种批次的迭代更容易。

下面开始训练模型，代码如下：

```python
# We use the pre-defined number of epochs to determine how many iterations to train the network on
for epoch in range(num_epochs):
    #Load in the data in batches using the train_loader object
    for i, (images, labels) in enumerate(train_loader):
        # Move tensors to the configured device
        images = images.to(device)
        labels = labels.to(device)
        
        # Forward pass
        outputs = model(images)
        loss = criterion(outputs, labels)
        
        # Backward and optimize
        optimizer.zero_grad()
        loss.backward()
        optimizer.step()
    
    print('Epoch [{}/{}], Loss: {:.4f}'.format(epoch+1, num_epochs, loss.item()))
```

看看代码的作用：

①首先迭代回合的数量，然后迭代训练数据中的批次。
②根据使用的设备（即 GPU 或 CPU）转换图像和标签。
③在前向通行中，使用模型进行预测，并根据这些预测和实际标签计算损失。
④进行反向传递，实际更新权重以改进模型。
⑤在每次更新之前，使用 optimizer.zero_grad() 函数将梯度设置为零。
⑥使用 loss.backward() 函数计算新的梯度。
⑦使用 optimizer.step() 函数更新权重。

输出如下：

```
Epoch [1/20], Loss: 1.5199
Epoch [2/20], Loss: 1.7072
Epoch [3/20], Loss: 1.7464
Epoch [4/20], Loss: 0.9564
Epoch [5/20], Loss: 1.6862
Epoch [6/20], Loss: 1.1571
Epoch [7/20], Loss: 0.9546
Epoch [8/20], Loss: 1.1339
Epoch [9/20], Loss: 1.4111
Epoch [10/20], Loss: 1.3795
Epoch [11/20], Loss: 0.3771
Epoch [12/20], Loss: 0.6321
Epoch [13/20], Loss: 1.0079
Epoch [14/20], Loss: 0.8751
Epoch [15/20], Loss: 0.5534
Epoch [16/20], Loss: 0.5643
Epoch [17/20], Loss: 0.2529
Epoch [18/20], Loss: 0.4387
Epoch [19/20], Loss: 0.3719
Epoch [20/20], Loss: 0.3089
```

随着回合的增多，损失略有减少，这是一个好迹象。在最后是波动的，这可能意味着模型过拟合或 batch_size 很小。这时，不得不进行测试以了解发生了什么。

下面测试模型。测试代码与训练没有太大区别，除了计算梯度，没有更新任何权重，代码如下：

```
with torch.no_grad():
    correct = 0
    total = 0
    for images, labels in train_loader:
        images = images.to(device)
        labels = labels.to(device)
        outputs = model(images)
        _, predicted = torch.max(outputs.data, 1)
        total += labels.size(0)
        correct += (predicted == labels).sum().item()

    print('Accuracy of the network on the {} train images: {} %'.format(50000, 100 * correct / total))
```

这里将代码封装在 torch.no_grad() 中，因为不需要计算任何梯度。然后，使用模型预测每个批次，并计算它正确预测的数量。得到的最终结果准确率约为 83%。

```
Accuracy of the network on the 50000 train images: 83.36 %
```

3.3 PyTorch 基础知识

本节涉及深度学习的基本构建块：张量。

导入 PyTorch 并检查正在使用的版本，代码如下：

```
>>> import torch
>>> torch.__version__
```

输出如下：

```
'2.2.1+cpu'
```

张量的工作是用数字的方式表示数据。例如，可以将图像表示为具有形状 [3, 224, 224] 的张量，这意味着 [color_channels, height, width]，因为在图像中有 3 个颜色通道（red, green, blue），高度为 224 像素，宽度为 224 像素。在张量语言（用于描述张量的语言）中，张量有 3 个维度，分别是颜色通道、高度和宽度。

下面通过编写代码来了解更多关于张量的信息。

3.3.1 创建张量

首先创建的是标量。标量是一个单一的数，又称零维张量。

```
>>> # Scalar
>>> scalar = torch.tensor(7)
>>> scalar
tensor(7)
```

可以使用 ndim 属性来检查张量的维数，代码如下：

```
>>> scalar.ndim
0
```

怎么从张量中检索数字呢？可以使用 item() 方法，代码如下：

```
>>> # Get the Python number within a tensor (only works with one-element tensors)
>>> scalar.item()
7
```

向量是一维张量，可以包含许多数字。

可以用一个向量 [3,2] 来描述房子里的 [卧室、浴室]。也可以用 [3,2,2] 来描述房子里的 [卧室、浴室、停车场]。

向量在其所能表示的内容上是灵活的。

```
>>> # Vector
>>> vector = torch.tensor([7, 7])
```

```
>>> vector
tensor([7, 7])
```

你认为它会有多少个维度?

```
>>> # Check the number of dimensions of vector
>>> vector.ndim
1
```

可以通过外部方括号([ ])的数量来判断 PyTorch 中张量的维数。

张量的另一个重要概念是它们的形状属性。形状告诉它们内部的元素是如何排列的。让我们看看向量的形状,代码如下:

```
>>> # Check shape of vector
>>> vector.shape
torch.Size([2])
```

上述代码返回 torch.Size([2]),这意味着向量具有 [2] 的形状。这是因为在方括号内放置了两个元素([7, 7])。

现在看一个矩阵,代码如下:

```
>>> # Matrix
>>> MATRIX = torch.tensor([[7, 8],
...                        [9, 10]])
>>> MATRIX
tensor([[ 7,  8],
        [ 9, 10]])
```

矩阵和向量一样灵活,只是它们有一个额外的维度,代码如下:

```
>>> # Check number of dimensions
>>> MATRIX.ndim
2
```

矩阵有两个维度。你认为它会是什么形状?

```
>>> MATRIX.shape
torch.Size([2, 2])
```

得到了 torch.Size([2, 2]),因为 MATRIX 是两个元素深和两个元素宽。

创建一个张量。

```
>>> # Tensor
>>> TENSOR = torch.tensor([[[1, 2, 3],
...                         [3, 6, 9],
...                         [2, 4, 5]]])
>>> TENSOR
tensor([[[1, 2, 3],
```

```
        [3, 6, 9],
        [2, 4, 5]]])
```

你认为它有多少维度？

```
>>> # Check number of dimensions for TENSOR
>>> TENSOR.ndim
3
```

检查它的形状。

```
>>> # Check shape of TENSOR
>>> TENSOR.shape
torch.Size([1, 3, 3])
```

维度从外到内。这意味着有一个 3×3 的 1 个维度。

3.3.2 随机张量

我们已经建立了张量来表示某种形式的数据。像神经网络这样的机器学习模型在张量中操纵和寻找模式。但是，当使用 PyTorch 构建机器学习模型时，很少会手动创建张量。相反，机器学习模型通常从大的随机数张量开始，并在处理数据时调整这些随机数，以更好地表示数据。

作为一名数据科学家，可以定义机器学习模型如何启动、查看数据和更新其随机数。

现在，看看如何创建一个随机数张量。可以使用 torch.rand() 并传入 size 参数来完成此操作，代码如下：

```
>>> # Create a random tensor of size (3, 4)
>>> random_tensor = torch.rand(size=(3, 4))
>>> random_tensor, random_tensor.dtype
(tensor([[0.3631, 0.3493, 0.7014, 0.8026],
        [0.7313, 0.7560, 0.5939, 0.4873],
        [0.9999, 0.1685, 0.6585, 0.3015]]), torch.float32)
```

torch.rand() 的灵活性在于，可以将大小调整为我们想要的任何大小。例如，想要一个通用图像形状为 [224, 224, 3]([高度，宽度，颜色通道]) 的随机张量，代码如下：

```
>>> # Create a random tensor of size (224, 224, 3)
>>> random_image_size_tensor = torch.rand(size=(224, 224, 3))
>>> random_image_size_tensor.shape, random_image_size_tensor.ndim
(torch.Size([224, 224, 3]), 3)
```

3.3.3 零和一

有时只想用零或一来填充张量。这种情况在做掩码时经常发生。

用 torch.zeros() 可以创建一个全是零的张量,代码如下:

```
>>> # Create a tensor of all zeros
>>> zeros = torch.zeros(size=(3, 4))
>>> zeros, zeros.dtype
(tensor([[0., 0., 0., 0.],
        [0., 0., 0., 0.],
        [0., 0., 0., 0.]]), torch.float32)
```

可以使用 torch.ones() 创建一个全是 1 的张量,代码如下:

```
>>> # Create a tensor of all ones
>>> ones = torch.ones(size=(3, 4))
>>> ones, ones.dtype# Create a tensor of all ones
(tensor([[1., 1., 1., 1.],
        [1., 1., 1., 1.],
        [1., 1., 1., 1.]]), torch.float32)
```

3.3.4 范围张量

有时可能需要一个数字范围,如 1 ~ 10 或 0 ~ 100。这时可以用 torch.arange(start, end, step)。

- start= 范围的开始(如 0)。
- end= 范围结束(如 10)。
- step= 每个值之间的步长(如 1)。

```
>>> # Create a range of values 0 to 10
>>> zero_to_ten = torch.arange(start=0, end=10, step=1)
>>> zero_to_ten
tensor([0, 1, 2, 3, 4, 5, 6, 7, 8, 9])
```

有时,可能想要一个特定类型的张量与另一个张量具有相同的形状。例如,一个全零的张量,其形状与前一个张量相同。要做到这一点,可以使用 torch.zeros_like(input) 或 torch.ones_like(input),它们分别返回一个填充有与输入形状相同的 0 或 1 的张量,代码如下:

```
>>> # Can also create a tensor of zeros similar to another tensor
>>> ten_zeros = torch.zeros_like(input=zero_to_ten) # will have same shape
>>> ten_zeros
tensor([0, 0, 0, 0, 0, 0, 0, 0, 0, 0])
```

3.3.5 张量数据类型

PyTorch 中有许多不同的张量数据类型。有些是专门针对 CPU 的，有些则更适合 GPU。

最常见的类型是 torch.foat32 或 torch.floot，这被称为 32 位浮点。也有 16 位浮点（torc.float16 或 torc.half）和 64 位浮点（torch.float64 或 torc.double）。

更令人困惑的是，还有 8 位、16 位、32 位和 64 位整数。

所有这些都与计算的精度有关。

精度是用来描述一个数字的细节量。精度值越高，用于表示数字的细节就越多，因此数据也就越多。

这在深度学习和数值计算中很重要，因为要做很多运算，如果需要计算的细节越多，则需要使用的计算就越多。因此，精度较低的数据类型通常计算速度较快，但在准确性等评估指标上牺牲了一些性能。

下面介绍如何创建一些具有特定数据类型的张量。可以使用 dtype 参数来完成此操作，代码如下：

```
>>> # Default datatype for tensors is float32
>>> float_32_tensor = torch.tensor([3.0, 6.0, 9.0],
...                                dtype=None,
...                                device=None,
...                                requires_grad=False)
>>>
>>> float_32_tensor.shape, float_32_tensor.dtype, float_32_tensor.device
(torch.Size([3]), torch.float32, device(type='cpu'))
```

除了形状问题（张量形状不匹配），在 PyTorch 中还会遇到另外两个最常见的问题，即数据类型和设备问题。例如，其中一个张量是 torch.float32，另一个是 torch.float16。或者一个张量在 CPU 上，另一个在 GPU 上。

下面创建一个 dtype=torch.foat16 的张量，代码如下：

```
>>> float_16_tensor = torch.tensor([3.0, 6.0, 9.0],
...                                dtype=torch.float16) # torch.half would also work
>>>
>>> float_16_tensor.dtype
torch.float16
```

3.3.6 从张量获取信息

张量的 3 个最常见的属性如下。

- shape——张量是什么形状？
- Dtype——张量中的元素存储在什么数据类型中？
- device——张量存储在什么设备上？

下面创建一个随机张量，并找出它的详细信息，代码如下：

```
>>> # Create a tensor
>>> some_tensor = torch.rand(3, 4)
>>>
>>> # Find out details about it
>>> print(some_tensor)
tensor([[2.7107e-01, 3.1900e-01, 1.7889e-02, 2.5988e-04],
        [6.9816e-01, 4.9283e-01, 8.6063e-01, 6.3694e-01],
        [2.1985e-01, 6.4265e-02, 4.8384e-01, 5.1195e-01]])
>>> print(f"Shape of tensor: {some_tensor.shape}")
Shape of tensor: torch.Size([3, 4])
>>> print(f"Datatype of tensor: {some_tensor.dtype}")
Datatype of tensor: torch.float32
>>> print(f"Device tensor is stored on: {some_tensor.device}") # will default to CPU
Device tensor is stored on: cpu
```

3.3.7 操纵张量

在深度学习中，数据（图像、文本、视频、音频、蛋白质结构等）被表示为张量。

模型通过研究这些张量并对张量执行一系列操作来学习，以创建输入数据中模式的表示。

这些操作通常是相加、相减、相乘、除、矩阵乘法。

当然，还有一些操作，但这些是神经网络的基本组成部分。

以正确的方式堆叠这些构建块，就可以创建最复杂的神经网络。

下面从一些基本运算开始介绍，如加法（+）、减法（-）、乘法（*），代码如下：

```
>>> # Create a tensor of values and add a number to it
>>> tensor = torch.tensor([1, 2, 3])
>>> tensor + 10
tensor([11, 12, 13])
>>> # Multiply it by 10
>>> tensor * 10
tensor([10, 20, 30])
```

注意，上面的张量值最终并不是 tensor([10, 20, 30])，这是因为张量内部的值不会改变，除非它们被重新分配，代码如下：

```
>>> # Tensors don't change unless reassigned
```

```
>>> tensor
tensor([1, 2, 3])
```

减去一个数字,重新分配张量变量,代码如下:

```
>>> # Subtract and reassign
>>> tensor = tensor - 10
>>> tensor
tensor([-9, -8, -7])
>>> # Add and reassign
>>> tensor = tensor + 10
>>> tensor
tensor([1, 2, 3])
```

PyTorch 还有一组内置函数,如 torch.mul() 和 torch.add(),用于执行基本运算,代码如下:

```
>>> # Can also use torch functions
>>> torch.multiply(tensor, 10)
tensor([10, 20, 30])
>>> # Original tensor is still unchanged
>>> tensor
tensor([1, 2, 3])
```

更常见的是使用像"*"这样的运算符符号,而不是 torch.mul(),代码如下:

```
>>> # Element-wise multiplication
>>> # (each element multiplies its equivalent, index 0->0, 1->1, 2->2)
>>> print(tensor, "*", tensor)
tensor([1, 2, 3]) * tensor([1, 2, 3])
>>> print("Equals:", tensor * tensor)
Equals: tensor([1, 4, 9])
```

机器学习和深度学习算法中最常见的操作之一是矩阵乘法。PyTorch 在 torch.matmul() 方法中可以实现矩阵乘法功能。

矩阵乘法的两个主要规则如下。

①内部尺寸必须匹配:

- (3, 2) @ (3, 2) 不起作用
- (2, 3) @ (3, 2) 将起作用
- (3, 2) @ (2, 3) 将起作用

②得到的矩阵具有外部尺寸的形状:

- (2, 3) @ (3, 2) -> (2, 2)
- (3, 2) @ (2, 3) -> (3, 3)

创建一个张量,并对其执行元素乘法和矩阵乘法,代码如下:

```
>>> import torch
>>> tensor = torch.tensor([1, 2, 3])
>>> tensor.shape
torch.Size([3])
>>> # Element-wise matrix multiplication
>>> tensor * tensor
tensor([1, 4, 9])
>>> # Matrix multiplication
>>> torch.matmul(tensor, tensor)
tensor(14)
>>> # Can also use the "@" symbol for matrix multiplication, though not recommended
>>> tensor @ tensor
tensor(14)
```

可以手工做矩阵乘法，但不建议这样做，因为内置的 torch.matmul() 方法速度更快。

3.3.8 深度学习中最常见的错误之一（形状错误）

因为很多深度学习都是对矩阵进行乘法和运算，而矩阵对可以组合的形状和大小有严格的规则，所以，在深度学习中最常见的错误之一是形状不匹配，示例代码如下：

```
>>> # Shapes need to be in the right way
>>> tensor_A = torch.tensor([[1, 2],
...                          [3, 4],
...                          [5, 6]], dtype=torch.float32)
>>>
>>> tensor_B = torch.tensor([[7, 10],
...                          [8, 11],
...                          [9, 12]], dtype=torch.float32)
>>>
>>> torch.matmul(tensor_A, tensor_B) # (this will error)
Traceback (most recent call last):
  File "<stdin>", line 1, in <module>
RuntimeError: mat1 and mat2 shapes cannot be multiplied (3x2 and 3x2)
```

可以通过使 tensor_A 和 tensor_B 的内部维度匹配来实现它们之间的矩阵乘法运算。其中一种方法是使用转置（切换给定张量的维度）。

可以使用以下任意一种方法在 PyTorch 中执行转置。

- torch.transpose(input, dim0, dim1)：其中 input 是要转置的张量，dim0 和 dim1 是要交换的维度。
- tensor.T：其中 tensor 是需要转置的张量。

让我们试试第二种方法，代码如下：

```
>>> # View tensor_A and tensor_B
>>> print(tensor_A)
tensor([[1., 2.],
        [3., 4.],
        [5., 6.]])
>>> print(tensor_B)
tensor([[ 7., 10.],
        [ 8., 11.],
        [ 9., 12.]])
>>> # View tensor_A and tensor_B.T
>>> print(tensor_A)
tensor([[1., 2.],
        [3., 4.],
        [5., 6.]])
>>> print(tensor_B.T)
tensor([[ 7.,  8.,  9.],
        [10., 11., 12.]])
>>> # The operation works when tensor_B is transposed
>>> print(f"Original shapes: tensor_A = {tensor_A.shape}, tensor_B = {tensor_B.shape}\n")
Original shapes: tensor_A = torch.Size([3, 2]), tensor_B = torch.Size([3, 2])

>>> print(f"New shapes: tensor_A = {tensor_A.shape} (same as above), tensor_B.T = {tensor_B.T.shape}\n")
New shapes: tensor_A = torch.Size([3, 2]) (same as above), tensor_B.T = torch.Size([2, 3])

>>> print(f"Multiplying: {tensor_A.shape} * {tensor_B.T.shape} <- inner dimensions match\n")
Multiplying: torch.Size([3, 2]) * torch.Size([2, 3]) <- inner dimensions match

>>> print("Output:\n")
Output:

>>> output = torch.matmul(tensor_A, tensor_B.T)
>>> print(output)
tensor([[ 27.,  30.,  33.],
        [ 61.,  68.,  75.],
        [ 95., 106., 117.]])
>>> print(f"\nOutput shape: {output.shape}")

Output shape: torch.Size([3, 3])
```

也可以使用 torch.mm()，它是 torch.matmul() 的缩写。

```
>>> torch.mm(tensor_A, tensor_B.T)
```

```
tensor([[ 27.,  30.,  33.],
        [ 61.,  68.,  75.],
        [ 95., 106., 117.]])
```

如果没有转置,矩阵乘法的规则就不满足,会得到如上所述的错误。

神经网络充满了矩阵乘法和点积。

torch.nn.Linear() 模块,也称为前馈层或全连接层,实现输入 x 和权重矩阵 A 之间的矩阵乘法。

$$y = x \cdot A^T + b$$

这里:

- x 是层的输入。
- A 是该层创建的权重矩阵,它最初是随机数,随着神经网络学习更好地表示数据中的模式,这些随机数会被调整。
- b 是用于稍微偏移权重和输入的偏置项。
- y 是输出。

尝试更改下面 in_features 和 out_features 的值,看看会发生什么。

```
>>> # Since the linear layer starts with a random weights matrix, let's make it reproducible
>>> torch.manual_seed(42)
<torch._C.Generator object at 0x000001F3FE670630>
>>> # This uses matrix multiplication
>>> linear = torch.nn.Linear(in_features=2, # in_features = matches inner dimension of input
...                          out_features=6) # out_features = describes outer value
>>> x = tensor_A
>>> output = linear(x)
>>> print(f"Input shape: {x.shape}\n")
Input shape: torch.Size([3, 2])

>>> print(f"Output:\n{output}\n\nOutput shape: {output.shape}")# Since the linear layer starts with a random weights matrix, let's make it reproducible (more on this later)
Output:
tensor([[2.2368, 1.2292, 0.4714, 0.3864, 0.1309, 0.9838],
        [4.4919, 2.1970, 0.4469, 0.5285, 0.3401, 2.4777],
        [6.7469, 3.1648, 0.4224, 0.6705, 0.5493, 3.9716]],
       grad_fn=<AddmmBackward0>)

Output shape: torch.Size([3, 6])
>>> torch.manual_seed(42)
<torch._C.Generator object at 0x000001F3FE670630>
>>> # This uses matrix multiplication
```

```
>>> linear = torch.nn.Linear(in_features=2, # in_features = matches inner dimension of input
...                          out_features=6) # out_features = describes outer value
>>> x = tensor_A
>>> output = linear(x)
>>> print(f"Input shape: {x.shape}\n")
Input shape: torch.Size([3, 2])

>>> print(f"Output:\n{output}\n\nOutput shape: {output.shape}")
Output:
tensor([[2.2368, 1.2292, 0.4714, 0.3864, 0.1309, 0.9838],
        [4.4919, 2.1970, 0.4469, 0.5285, 0.3401, 2.4777],
        [6.7469, 3.1648, 0.4224, 0.6705, 0.5493, 3.9716]],
       grad_fn=<AddmmBackward0>)

Output shape: torch.Size([3, 6])
```

如果以前从未做过,矩阵乘法一开始可能是一个令人困惑的话题。但当你用了几次,并且破解了一些神经网络后,便会注意到它无处不在。

3.3.9 求最小值、最大值、平均值、总和等

下面介绍聚合张量的方法。

首先将创建一个张量,然后找到它的最大值、最小值、平均值和总和,代码如下:

```
>>> # Create a tensor
>>> x = torch.arange(0, 100, 10)
>>> x
tensor([ 0, 10, 20, 30, 40, 50, 60, 70, 80, 90])
```

进行一些聚合,代码如下:

```
>>> print(f"Minimum: {x.min()}")
Minimum: 0
>>> print(f"Maximum: {x.max()}")
Maximum: 90
>>> # print(f"Mean: {x.mean()}") # this will error
>>> print(f"Mean: {x.type(torch.float32).mean()}") # won't work without float datatype
Mean: 45.0
>>> print(f"Sum: {x.sum()}")
Sum: 450
```

也可以使用 torch 方法执行与上述相同的操作,代码如下:

```
>>> torch.max(x), torch.min(x), torch.mean(x.type(torch.float32)), torch.sum(x)
(tensor(90), tensor(0), tensor(45.), tensor(450))
```

3.3.10 最大值、最小值的所处位置

可以找到张量的索引,其中最大值和最小值分别出现在 torch.argmax() 和 torch.argmin() 中。这在只想要最高(或最低)值所在的位置而不是实际值本身的情况下很有用,示例代码如下:

```
>>> # Create a tensor
>>> tensor = torch.arange(10, 100, 10)
>>> print(f"Tensor: {tensor}")
Tensor: tensor([10, 20, 30, 40, 50, 60, 70, 80, 90])
>>>
>>> # Returns index of max and min values
>>> print(f"Index where max value occurs: {tensor.argmax()}")
Index where max value occurs: 8
>>> print(f"Index where min value occurs: {tensor.argmin()}")
Index where min value occurs: 0
```

3.3.11 更改张量数据类型

如前所述,深度学习操作的一个常见问题是在不同的数据类型中使用张量。如果一个张量在 torch.float64 中,另一个张量在 torch.float32 中,则可能会遇到一些错误。解决办法如下:使用 torch.Tensor.type(dtype=None) 更改张量的数据类型,其中 dtype 参数是要使用的数据类型。

首先,创建一个张量并检查它的数据类型,代码如下:

```
>>> # Create a tensor and check its datatype
>>> tensor = torch.arange(10., 100., 10.)
>>> tensor.dtype
torch.float32
```

然后,创建另一个与以前相同的张量,但将其数据类型更改为 torch.float16,代码如下:

```
>>> # Create a float16 tensor
>>> tensor_float16 = tensor.type(torch.float16)
>>> tensor_float16
tensor([10., 20., 30., 40., 50., 60., 70., 80., 90.], dtype=torch.float16)
```

最后,创建一个 torch.int8 类型的张量,代码如下:

```
>>> # Create a int8 tensor
>>> tensor_int8 = tensor.type(torch.int8)
>>> tensor_int8
tensor([10, 20, 30, 40, 50, 60, 70, 80, 90], dtype=torch.int8)
```

3.3.12 重塑、堆叠、压缩和解压

通常情况下，用户会想重塑或改变张量的尺寸，而不想实际改变张量内部的值。一些流行的方法如表 3-1 所示。

表 3-1 重塑、堆叠、压缩和解压方法

方　　法	描　　述
torch.reshape(input, shape)	重塑输入的形状，也可以使用 torch.Tensor.reshape()
Tensor.view(shape)	以不同的形状返回原始张量的视图，但与原始张量共享相同的数据
torch.stack(tensors, dim=0)	沿新维度连接张量序列，所有张量的大小必须相同
torch.squeeze(input)	压缩输入以从张量中移除维度为 1 的维度
torch.unsqueeze(input, dim)	返回在 dim 处添加维度值为 1 的输入
torch.permute(input, dims)	返回原始输入的视图，将其尺寸变换为 dims

为什么要做这些？因为深度学习模型都是关于以某种方式操纵张量的。由于矩阵乘法的规则，如果有形状不匹配，会遇到错误。这些方法可以帮助用户确保张量中的正确元素与其他张量中的适当元素混合。

首先，创建一个张量，代码如下：

```
>>> # Create a tensor
>>> import torch
>>> x = torch.arange(1., 8.)
>>> x, x.shape
(tensor([1., 2., 3., 4., 5., 6., 7.]), torch.Size([7]))
```

使用 torch.reshape() 添加一个额外的维度，代码如下：

```
>>> # Add an extra dimension
>>> x_reshaped = x.reshape(1, 7)
>>> x_reshaped, x_reshaped.shape
(tensor([[1., 2., 3., 4., 5., 6., 7.]]), torch.Size([1, 7]))
```

也可以使用 torch.view() 更改视图，代码如下：

```
>>> # Change view (keeps same data as original but changes view)
>>> z = x.view(1, 7)
>>> z, z.shape
```

```
(tensor([[1., 2., 3., 4., 5., 6., 7.]]), torch.Size([1, 7]))
```

使用 torch.view() 更改张量的视图只会创建相同张量的新视图。因此，改变视图也会改变原始张量，代码如下：

```
>>> # Changing z changes x
>>> z[:, 0] = 5
>>> z, x
(tensor([[5., 2., 3., 4., 5., 6., 7.]]), tensor([5., 2., 3., 4., 5., 6., 7.]))
```

如果想把新的张量叠加在自己上面 5 次，可以用 torch.stack() 来实现，代码如下：

```
>>> # Stack tensors on top of each other
>>> x_stacked = torch.stack([x, x, x, x], dim=0)
>>> x_stacked
tensor([[5., 2., 3., 4., 5., 6., 7.],
        [5., 2., 3., 4., 5., 6., 7.],
        [5., 2., 3., 4., 5., 6., 7.],
        [5., 2., 3., 4., 5., 6., 7.]])
```

如何从张量中移除所有单个维度？可以使用 torch.squeeze()，代码如下：

```
>>> print(f"Previous tensor: {x_reshaped}")
Previous tensor: tensor([[5., 2., 3., 4., 5., 6., 7.]])
>>> print(f"Previous shape: {x_reshaped.shape}")
Previous shape: torch.Size([1, 7])
>>>
>>> # Remove extra dimension from x_reshaped
>>> x_squeezed = x_reshaped.squeeze()
>>> print(f"\nNew tensor: {x_squeezed}")

New tensor: tensor([5., 2., 3., 4., 5., 6., 7.])
>>> print(f"New shape: {x_squeezed.shape}")
New shape: torch.Size([7])
```

要执行 torch.squeeze() 的相反操作，可以使用 torch.unsqueeze() 在特定索引处添加维度值 1，代码如下：

```
>>> print(f"Previous tensor: {x_squeezed}")
Previous tensor: tensor([5., 2., 3., 4., 5., 6., 7.])
>>> print(f"Previous shape: {x_squeezed.shape}")
Previous shape: torch.Size([7])
>>>
>>> ## Add an extra dimension with unsqueeze
>>> x_unsqueezed = x_squeezed.unsqueeze(dim=0)
>>> print(f"\nNew tensor: {x_unsqueezed}")
```

```
New tensor: tensor([[5., 2., 3., 4., 5., 6., 7.]])
>>> print(f"New shape: {x_unsqueezed.shape}")
New shape: torch.Size([1, 7])
```

也可以使用 torch.permute(input, dims) 重新排列轴值的顺序，代码如下：

```
>>> # Create tensor with specific shape
>>> x_original = torch.rand(size=(224, 224, 3))
>>>
>>> # Permute the original tensor to rearrange the axis order
>>> x_permuted = x_original.permute(2, 0, 1) # shifts axis 0->1, 1->2, 2->0
>>>
>>> print(f"Previous shape: {x_original.shape}")
Previous shape: torch.Size([224, 224, 3])
>>> print(f"New shape: {x_permuted.shape}")
New shape: torch.Size([3, 224, 224])
```

3.3.13 索引（从张量中选择数据）

有时，需要从张量中选择特定的数据（例如，仅第一列或第二行）。可以使用索引。如果曾经在 Python 列表或 NumPy 数组上进行过索引，那么在 PyTorch 中使用张量进行索引是非常相似的，代码如下：

```
>>> # Create a tensor
>>> import torch
>>> x = torch.arange(1, 10).reshape(1, 3, 3)
>>> x, x.shape
(tensor([[[1, 2, 3],
         [4, 5, 6],
         [7, 8, 9]]]), torch.Size([1, 3, 3]))
```

索引值为外部维度 -> 内部维度（检查方括号）。

```
>>> # Let's index bracket by bracket
>>> print(f"First square bracket:\n{x[0]}")
First square bracket:
tensor([[1, 2, 3],
        [4, 5, 6],
        [7, 8, 9]])
>>> print(f"Second square bracket: {x[0][0]}")
Second square bracket: tensor([1, 2, 3])
>>> print(f"Third square bracket: {x[0][0][0]}")
Third square bracket: 1
```

这可以使用":"指定"此维度中的所有值"，然后使用逗号，添加另一个维度，代码如下：

```
>>> # Get all values of 0th dimension and the 0 index of 1st dimension
>>> x[:, 0]
tensor([[1, 2, 3]])
>>> # Get all values of 0th & 1st dimensions but only index 1 of 2nd dimension
>>> x[:, :, 1]
tensor([[2, 5, 8]])
>>> # Get all values of the 0 dimension but only the 1 index value of the 1st and 2nd dimension
>>> x[:, 1, 1]
tensor([5])
>>> # Get index 0 of 0th and 1st dimension and all values of 2nd dimension
>>> x[0, 0, :] # same as x[0][0]
tensor([1, 2, 3])
```

3.3.14 PyTorch 张量和 NumPy

NumPy 是一个流行的 Python 数值计算库，PyTorch 具有与之良好交互的功能。

NumPy 到 PyTorch（及返回）的两种主要方法如下：

- torch.from_numpy(ndarray) – NumPy 数组 -> PyTorch 张量。
- torch.Tensor.numpy() — PyTorch 张量 -> NumPy 数组。

```
>>> # NumPy array to tensor
>>> import torch
>>> import numpy as np
>>> array = np.arange(1.0, 8.0)
>>> tensor = torch.from_numpy(array)
>>> array, tensor
(array([1., 2., 3., 4., 5., 6., 7.]), tensor([1., 2., 3., 4., 5., 6., 7.], dtype=torch.float64))
```

默认情况下，NumPy 数组是使用数据类型 float64 创建的，如果将其转换为 PyTorch 张量，它将保持相同的数据类型。但是，许多 PyTorch 计算默认使用 float32 类型。因此，如果想转换 NumPy 数组 (float64)->PyTorch 张量 (float64)->PyTorch 张量 (float32)，可以使用 tensor = torch.from_numpy(array).type(torch.float32)。

如果想从 PyTorch 张量到 NumPy 数组，可以调用 tensor.numpy()，代码如下：

```
>>> # Tensor to NumPy array
>>> tensor = torch.ones(7) # create a tensor of ones with dtype=float32
>>> numpy_tensor = tensor.numpy() # will be dtype=float32 unless changed
```

```
>>> tensor, numpy_tensor
(tensor([1., 1., 1., 1., 1., 1., 1.]), array([1., 1., 1., 1., 1., 1., 1.],
dtype=float32))
```

如果更改原始张量，则新的 numpy_tensor 保持不变，代码如下：

```
>>> # Change the tensor, keep the array the same
>>> tensor = tensor + 1
>>> tensor, numpy_tensor
(tensor([2., 2., 2., 2., 2., 2., 2.]), array([1., 1., 1., 1., 1., 1., 1.],
dtype=float32))
```

3.3.15 再现性（试图从随机中提取随机性）

随着对神经网络和机器学习的了解越来越多，会发现随机性在多大程度上起作用。就是伪随机性。因为毕竟计算机从根本上是确定性的，所以，它们产生的随机性是模拟的随机性。那么，这与神经网络和深度学习有什么关系呢？神经网络从随机数开始描述数据中的模式，并试图使用张量运算来改进这些随机数，以更好地描述数据中的模式。

简言之：从随机数开始 -> 张量运算 -> 努力做得更好。

尽管随机性很好，功能强大，但有时，希望随机性少一点。为什么？因为这样可以执行可重复的实验。例如，可以创建一个能够实现 X 性能的算法。然后你的朋友试了试，验证你没有疯。他们怎么能做出这种事呢？这就是再现性的作用所在。换句话说，你能否在你的计算机上运行与我相同的代码得到相同（或非常相似）的结果？让我们看一个 PyTorch 中再现性的简短示例。我们首先创建两个随机张量，因为它们是随机的，你会期望它们不同，对吧？

```
>>> import torch
>>>
>>> # Create two random tensors
>>> random_tensor_A = torch.rand(3, 4)
>>> random_tensor_B = torch.rand(3, 4)
>>>
>>> print(f"Tensor A:\n{random_tensor_A}\n")
Tensor A:
tensor([[0.8016, 0.3649, 0.6286, 0.9663],
        [0.7687, 0.4566, 0.5745, 0.9200],
        [0.3230, 0.8613, 0.0919, 0.3102]])

>>> print(f"Tensor B:\n{random_tensor_B}\n")
Tensor B:
tensor([[0.9536, 0.6002, 0.0351, 0.6826],
        [0.3743, 0.5220, 0.1336, 0.9666],
```

```
        [0.9754, 0.8474, 0.8988, 0.1105]])
>>> print(f"Does Tensor A equal Tensor B? (anywhere)")
Does Tensor A equal Tensor B? (anywhere)
>>> random_tensor_A == random_tensor_B
tensor([[False, False, False, False],
        [False, False, False, False],
        [False, False, False, False]])
```

正如你所料,张量的值是不同的。但是,如果想创建两个具有相同值的随机张量,该怎么办?张量仍将包含随机值,但它们将具有相同的味道。这就是torch.manual_seed(seed) 的作用所在,其中 seed 是一个整数(如42,它可以是任何值),可以给随机性增加味道。

下面通过创建一些更有味道的随机张量来尝试一下。

```
>>> import torch
>>> import random
>>>
>>> # # Set the random seed
>>> RANDOM_SEED=42
>>> torch.manual_seed(seed=RANDOM_SEED)
<torch._C.Generator object at 0x000001F3FE670630>
>>> random_tensor_C = torch.rand(3, 4)
>>>
>>> # Have to reset the seed every time a new rand() is called
>>> # Without this, tensor_D would be different to tensor_C
>>> torch.random.manual_seed(seed=RANDOM_SEED)
<torch._C.Generator object at 0x000001F3FE670630>
>>> random_tensor_D = torch.rand(3, 4)
>>>
>>> print(f"Tensor C:\n{random_tensor_C}\n")
Tensor C:
tensor([[0.8823, 0.9150, 0.3829, 0.9593],
        [0.3904, 0.6009, 0.2566, 0.7936],
        [0.9408, 0.1332, 0.9346, 0.5936]])

>>> print(f"Tensor D:\n{random_tensor_D}\n")
Tensor D:
tensor([[0.8823, 0.9150, 0.3829, 0.9593],
        [0.3904, 0.6009, 0.2566, 0.7936],
        [0.9408, 0.1332, 0.9346, 0.5936]])

>>> print(f"Does Tensor C equal Tensor D? (anywhere)")
Does Tensor C equal Tensor D? (anywhere)
```

```
>>> random_tensor_C == random_tensor_D
tensor([[True, True, True, True],
        [True, True, True, True],
        [True, True, True, True]])
```

很好！这看起来像是播种成功了。

3.4　transformer 架构

transformer 架构代码演练单前向传播。

导入一些包：

```
import torch
import torch.nn as nn
import torch.nn.functional as F
import math
import numpy as np
import matplotlib.pyplot as plt
```

定义超参数：

```
d_embed = 512        # embedding size for the attention modules
num_heads = 8        # Number of attention heads
num_batches = 1      # number of batches (1 makes it easier to see what is going on)
vocab = 50000        # vocab size
max_len = 5000       # Max length of TODO what exactly?
n_layers = 1         # number of attention layers (not used but would be an expected hyper-parameter)
d_ff = 2048          # hidden state size in the feed forward layers
epsilon = 1e-6       # epsilon to use when we need a small non-zero number
```

在这里，创建一些虚拟输入数据，共由 3 个表征组成。第二个表征将是屏蔽表征。最初，有一个大小为 batch_size×sequence_length 的输入 x。自始至终，将使用 x 表示源自输入序列的张量，使用 y 表示源自目标序列的张量，代码如下：

```
x = torch.tensor([[1, 2, 3]])      # input will be 3 tokens
y = torch.tensor([[1, 2, 3]])      # target will be same as the input for many applications
x_mask = torch.tensor([[1, 0, 1]]) # Mask the 2nd input token
y_mask = torch.tensor([[1, 0, 1]]) # Mask the 2nd target token
print("x", x.size())
print("y", y.size())
```

3.4.1 编码器

本节展示了编码器中一个注意力层的演练。编码器的目的是创建一个隐藏状态,即输入序列的编码表示,然后将隐藏状态传递给解码器。

编码器嵌入使用传统的嵌入层将表征转换为大小为 d_embed 的嵌入。然后通过 sqrt(d_model) 对嵌入激活进行缩放,以使其更大。结果是大小为 batch_size × sequence_length × embedding_size 的张量。

```
# Make the embedding module. It understands that each token should result in a
separate embedding.
emb = nn.Embedding(vocab, d_embed)
x = emb(x)
# Scale the embedding
x = x * math.sqrt(d_embed)
print(x.size())
```

接下来添加位置嵌入信息。下面的代码创建了一个叠加正弦和余弦波的模式,并将其添加到嵌入中。这根据嵌入的表征在序列中的位置来区分嵌入的表征。也就是说,如果一个输入序列有两个相同的表征,那么根据它们在序列中的位置,它们的嵌入最终会看起来有点不同。

```
# Start with an empty tensor
pe = torch.zeros(max_len, d_embed, requires_grad=False)
# array containing index values 0...max_len
position = torch.arange(0, max_len).unsqueeze(1)
divisor = torch.exp(torch.arange(0, d_embed, 2) * -(math.log(10000.0) / d_embed))
# Make overlapping sine and cosine wave inside positional embedding tensor
pe[:, 0::2] = torch.sin(position * divisor)
pe[:, 1::2] = torch.cos(position * divisor)
pe = pe.unsqueeze(0)
# Add the position embedding to the main embedding
x = x + pe[:, :x.size(1)]
print(x.size())
```

为了了解位置嵌入是如何工作的,可以在嵌入的每个维度中可视化添加到每个嵌入中的值(这里只可视化前 8 个维度),代码如下:

```
plt.figure(figsize=(15, 5))    # Make a plot
d_embed_plot = 16    # for illustration purposes, set embedding dimensions = 16
pe_plot = torch.zeros(max_len, d_embed_plot, requires_grad=False) # positional
embedding tensor
position_plot = torch.arange(0, max_len).unsqueeze(1)
divisor_plot = torch.exp(torch.arange(0, d_embed_plot, 2) * -(math.log(10000.0) /
d_embed_plot))
```

```
pe_plot[:, 0::2] = torch.sin(position_plot * divisor_plot)
pe_plot[:, 1::2] = torch.cos(position_plot * divisor_plot)
pe_plot = pe_plot.unsqueeze(0)
# plot it
y_plot = torch.zeros(1, 50, d_embed_plot)
y_plot = pe_plot[:, :y_plot.size(1)]
plt.plot(np.arange(50), y_plot[0, :, 0:4].data.numpy())
plt.legend(["dim %d"%p for p in range(8)])
```

编码器注意子层将重复 N 次。此代码演练将只带我们了解一个。编码器注意层由一个自注意模块和一个前馈模块组成。

自注意和前馈是用残差包裹的。残差连接将块的输入与块的输出相加。因此，可以将该块视为试图学习如何从输入中进行加法或减法。这为训练提供了稳定性，因为该块并不为向前和向后传播中发生的一切负全部责任。看看 transformer 的编码器，可以看到绕过自注意的剩余连接提供了与隐藏状态的直接连接。也就是说，底部的嵌入可以选择在最终隐藏状态编码方面做很多繁重的工作。自注意和其他子层可能会给最终的隐藏状态增加一点，如果有助于减少损失的话，可能会增加很多。关于残差的另一种思考方式就像传统计算机程序中计算一些副作用的子程序。一个子例程计算最终的隐藏状态，另一个子例程分支并计算自注意。但因为每个模块都必须在一个梯度路径上，所以，副例程必须对最终的损失有所贡献。

残差将输入重新添加到输出中，因此，中间发生的事情可以被认为是计算原始值的增量。

通常，不需要执行 clone() 来创建残差，但在每个步骤中都使用相同的 x 变量，因此克隆确保我们不会覆盖，代码如下：

```
x_residual = x.clone()
print(x.size())
```

在计算自注意之前，执行层规范化。层规范化通过减少值开始走向极端的机会来稳定训练。这是通过将所有值相对于平均值居中来实现的，代码如下：

```
mean = x.mean(-1, keepdim=True)
std = x.std(-1, keepdim=True)
W1 = nn.Parameter(torch.ones(d_embed))
b1 = nn.Parameter(torch.zeros(d_embed))
x = W1 * (x - mean) / (std + epsilon) + b1
print(x.size())
```

自注意是一个生成分数的过程，该分数指示每个表征与其他表征的关系。$0 \sim 1$ 的值的 seq_length × seq_length 矩阵中的每个值指示第 i 个表征对第 j 个表征的重要性。模型必须学习如何生成分数。

理解自注意的隐喻是一个哈希表。在哈希表中有一个键列表，每个键都与一个值相关联。一个查询被发送到哈希表，哈希表必须找到匹配的键并返回关联的值。假设这个哈希表是一个模糊哈希表，因为查询不必匹配任何键，所以，哈希表将返回看起来最接近查询的内容。

要注意的输入是 batch_size × sequence_length × embedding_size 矩阵。忽略批处理维度，我们得到的是一系列嵌入的表征。要注意将这个输入 x 复制 3 次，并称之为"查询"（q）、"键"（k）和"值"（v）。每个矩阵都经过一个线性层。这个线性层是网络学习得分的地方。它使每个矩阵都不同，如果它找到正确的、不同的矩阵，就会获得很好的注意力得分。如果它得到了很好的注意力分数，损失就会减少。

注意力得分以如下方法生成：首先，将 q 和 k 矩阵分解为多个部分（称为"头"）。这就是所谓的多头注意力。这样做的原因是为了让每个头部都能独立产生不同的注意力得分。这允许每个表征都有几个最好的其他表征。在实现中，只需将每个表征的块嵌入到不同的头中。

q 和 k 张量相乘。这将创建一个 batch_size × num_heads × sequence_length × sequence_length 矩阵。忽略批处理和头，可以将此矩阵解释为包含原始分数，其中每个单元格计算第 i 个表征与第 j 个表征的相关性（i 是行，j 是列）。

接下来，将这个矩阵通过 softmax 层。softmax 的秘密在于它可以像 argmax 一样——可以选择最佳匹配。softmax 将沿特定维度的所有值压缩为 0…1。但它真正做的是试图迫使一个特定的单元格的数字接近 1，其余的都接近 0。如果将这个 softmaxed 分数矩阵乘以 v 矩阵，本质上是在问（对于每个头）哪一列对每一行最好。回想一下，行和列对应于表征。所以我们在问，哪个表征与其他表征最匹配。同样，如果早期的线性层的参数正确，这种乘法将做出良好的选择，并且损失将得到改善。

在这一点上，可以将与 v 相乘的分数视为试图将除最相关的表征嵌入之外的所有内容归零（因为有多个头，所以有几个）。为了保持一致性，将把结果存储到 x 中，主要是参与度最高的表征嵌入（因为有多个头，所以有几个），再加上一点其他嵌入的表征，因为我们不能做实际的 argmax——我们能做的最好的事是让所有不相关的东西都接近零，这样它就不会影响其他任何东西。

这种分数与 v 矩阵的乘积就是我们所说的自注意。自注意本质上是一个点积，具有基本的学习评分函数。它告诉我们应该在哪里寻找好的信息。解码器稍后将使用它。

```
# Make three versions of x, for the query, key, and value
# We don't need to clone because these will immediately go through linear layers,
making new tensors
k = x # key
q = x # query
v = x # value
# Make three linear layers
# This is where the network learns to make scores
linear_k = nn.Linear(d_embed, d_embed)
```

```
    linear_q = nn.Linear(d_embed, d_embed)
    linear_v = nn.Linear(d_embed, d_embed)
    # We are going to fold the embedding dimensions and treat each fold as an
attention head
    d_k = d_embed // num_heads
    # Pass q, k, v through their linear layers
    q = linear_q(q)
    k = linear_k(k)
    v = linear_v(v)
    # Do the fold, treating each h dimenssions as a head
    # Put the head in the second position
    q = q.view(num_batches, -1, num_heads, d_k).transpose(1, 2)
    k = k.view(num_batches, -1, num_heads, d_k).transpose(1, 2)
    v = v.view(num_batches, -1, num_heads, d_k).transpose(1, 2)
    print("q", q.size())
    print("x", k.size())
    print("v", v.size())
```

为了产生注意力得分，将 q 和 k 相乘（并归一化）。我们需要应用掩码，这样掩码表征就不会注意自己。应用 softmax 来模拟 argmax（好的东西接近 1，不相关的东西接近 0）。如果你看 attn，你不会看到这种情况发生，因为线性层还没有经过训练。注意力得分最终应用于 v。

```
    d_k = q.size(-1)
    # Compute the raw scores by multiplying k and q (and normalize)
    scores = torch.matmul(k, q.transpose(-2, -1)) / math.sqrt(d_k)
    # Mask out the scores
    scores = scores.masked_fill(x_mask == 0, -epsilon)
    # Softmax the scores, ideally creating one score close to 1 and the rest close to 0
    # (Note: this won't happen if you look at the numbers because the linear layers haven't
    # learned anything yet.)
    attn = F.softmax(scores, dim = -1)
    print("attention", attn.size())
    # Apply the scores to v
    x = torch.matmul(attn, v)
    print("x", x.size())
```

以下是要注意的作用说明。在每个行和列下面的 attn 矩阵中，表示输入序列中的不同位置，使得 attn[i][j] 是第 i 个位置对第 j 个位置的亲和度。在一个完美的世界中，softmax 将每行中的一个元素推到接近 1，而其他元素则推到接近 0。将 attn 乘以 v，为每个位置选择一个嵌入（隐藏状态）。

```
    # Make fake attention scores with extreme values
    attn = torch.zeros(3, 3)
    attn[0,1] = 1
```

```
attn[1,2] = 1
attn[2,0] = 1
print("attn:")
print(attn)
# Make a fake v embedding
v = torch.tensor(list(map(lambda x:list(range(x*10,(x*10)+10)), list(range(3))))).float()
print("v:")
print(v)
print("Matmul result:")
print(torch.matmul(attn, v))
```

但现在嵌入都被交换了,很多位置中的一点点可以混合在一起。这就是为什么残差很重要,因为不能在它们的原始位置丢失原始嵌入。

重新组合多个注意力头(展开),代码如下:

```
x = x.transpose(1, 2).contiguous().view(num_batches, -1, num_heads * (d_embed // num_heads))
print(x.size())
```

从这一点来看,有一些表征嵌入被推向 1,一些表征嵌入则被推向 0。需要准备这个矩阵,将其重新添加到残差中。也就是说,无论这个转换产生什么,都是一组值,这些值将每个表征的原始嵌入值向上或向下更改一些增量,代码如下:

```
ff = nn.Linear(d_embed, d_embed)
x = ff(x)
print(x.size())
```

添加回残差,代码如下:

```
x = x_residual + x
print(x.size())
```

前馈模块是一个直接的解码和重新编码的嵌入加上自注意。在编码阶段结束时,我们想要的是一个隐藏状态。就像在序列到序列网络中一样,我们想要一堆隐藏状态,每个表征一个。这样,解码器将能够通过只查看该堆栈而不是迭代所有输入表征来回顾和关注对解码最有用的隐藏状态。因此,每个表征位置中的任何内容都必须代表输入中正在发生的事情。为了将矩阵移动到隐藏状态,我们扩展嵌入,给网络一些容量,然后再次将其折叠,迫使其进行权衡。

留出残差,代码如下:

```
x_residual = x.clone()
print(x.size())
```

预前馈层归一化,代码如下:

```
mean = x.mean(-1, keepdim=True)
```

```
std = x.std(-1, keepdim=True)
W2 = nn.Parameter(torch.ones(d_embed))
b2 = nn.Parameter(torch.zeros(d_embed))
x = W2 * (x - mean) / (std + epsilon) + b2
print(x.size())
```

这个前馈模块让嵌入增长,然后再次压缩它。这是将自注意模块的输出转换为隐藏状态编码过程的一部分,代码如下:

```
linear_expand = nn.Linear(d_embed, d_ff)
linear_compress = nn.Linear(d_ff, d_embed)
x = linear_compress(F.relu(linear_expand(x)))
print(x.size())
```

添加回残差,代码如下:

```
x = x_residual + x
print(x.size())
```

在重复自关注和前馈子层 N 次之后,应用最后一层归一化,代码如下:

```
mean = x.mean(-1, keepdim=True)
std = x.std(-1, keepdim=True)
Wn = nn.Parameter(torch.ones(d_embed))
bn = nn.Parameter(torch.zeros(d_embed))
x = Wn * (x - mean) / (std + epsilon) + bn
print(x.size())
```

在这一点上,我们应该有一个矩阵,存储在 x 中,可以将其解释为隐藏状态的堆栈。解码器将尝试关注此堆栈。

```
# Signify that the output is the hidden state
hidden = x
print(hidden.size())
```

3.4.2 解码器

解码器的工作原理与编码器非常相似,只是有一个主要的变化。除了自注意和前馈模块外,解码器还包括源注意模块,其中它关注编码器的隐藏状态输出。

我们将对 y 进行操作,y 是目标表征的序列。将目标视为输入似乎很奇怪。最接近的模拟是序列到序列网络,它将生成一系列输出表征,每次一个,与目标序列进行比较以计算损失。但在这里,我们不需要生成输出序列,因为没有递归,所以,只需要取目标输出,并将其视为 transformer 产生的输出。例外情况是屏蔽输出表征(通常与屏蔽输入的位置相同)。为了计算损失,我们只关心是否对屏蔽的目标表征进行了良好的预测。

解码器嵌入，代码如下：

```
emb_d = nn.Embedding(vocab, d_embed)
y = emb_d(y) * math.sqrt(d_embed)
print(y.size())
```

添加位置嵌入，代码如下：

```
pe = torch.zeros(max_len, d_embed, requires_grad=False)
position = torch.arange(0, max_len).unsqueeze(1)
divisor = torch.exp(torch.arange(0, d_embed, 2) * -(math.log(10000.0) / d_embed))
pe[:, 0::2] = torch.sin(position * divisor)
pe[:, 1::2] = torch.cos(position * divisor)
pe = pe.unsqueeze(0)
y = y + pe[:, :y.size(1)]
print(y.size())
```

解码器层将被重复 N 次。此代码演练将只带我们了解一个。解码器注意力层由自注意力、源注意力和前馈组成，每个都被残差包裹着。

留出残差，代码如下：

```
y_residual = y.clone()
print(y.size())
```

预自注意层规范化，代码如下：

```
mean = y.mean(-1, keepdim=True)
std = y.std(-1, keepdim=True)
W1_d = nn.Parameter(torch.ones(d_embed))
b1_d = nn.Parameter(torch.zeros(d_embed))
y = W1_d * (y - mean) / (std + epsilon) + b1_d
print(y.size())
```

自注意，代码如下：

```
k = y
q = y
v = y
linear_q_self = nn.Linear(d_embed, d_embed)
linear_k_self = nn.Linear(d_embed, d_embed)
linear_v_self = nn.Linear(d_embed, d_embed)
d_k = d_embed // num_heads
q = linear_q_self(q)
k = linear_k_self(k)
v = linear_v_self(v)
q = q.view(num_batches, -1, num_heads, d_k).transpose(1, 2)
k = k.view(num_batches, -1, num_heads, d_k).transpose(1, 2)
```

```python
v = v.view(num_batches, -1, num_heads, d_k).transpose(1, 2)
print("q", q.size())
print("k", k.size())
print("v", v.size())
d_k = q.size(-1)
scores = torch.matmul(k, q.transpose(-2, -1)) / math.sqrt(d_k)
scores = scores.masked_fill(y_mask == 0, -epsilon)
attn = F.softmax(scores, dim = -1)
print("attention", attn.size())
y = torch.matmul(attn, v)
print("y", y.size())
```

组装头，代码如下：

```python
y = y.transpose(1, 2).contiguous().view(num_batches, -1, num_heads * (d_embed // num_heads))
print(y.size())
```

后自注意前馈，代码如下：

```python
ff_d1 = nn.Linear(d_embed, d_embed)
y = ff_d1(y)
print(y.size())
```

添加回残差，代码如下：

```python
y = y_residual + y
print(y.size())
```

将残差放在一边，代码如下：

```python
y_residual = y.clone()
print(y.size())
```

源前注意层规范化，代码如下：

```python
mean = y.mean(-1, keepdim=True)
std = y.std(-1, keepdim=True)
W2_d = nn.Parameter(torch.ones(d_embed))
b2_d = nn.Parameter(torch.zeros(d_embed))
y = W2_d * (y - mean) / (std + epsilon) + b2_d
print(y.size())
```

源注意力的工作原理与自注意力一样，只是我们使用编码器的键和值来计算分数，并将其应用于解码器的查询。也就是说，基于编码器认为我们应该注意什么，我们实际上应该关注解码器序列的哪一部分。

```python
q = y
k = x # notice we are using x
```

```python
v = x # notice we are using x
linear_q_source = nn.Linear(d_embed, d_embed)
linear_k_source = nn.Linear(d_embed, d_embed)
linear_v_source = nn.Linear(d_embed, d_embed)
d_k = d_embed // num_heads
q = linear_q(q)
k = linear_k(k)
v = linear_v(v)
q = q.view(num_batches, -1, num_heads, d_k).transpose(1, 2)
k = k.view(num_batches, -1, num_heads, d_k).transpose(1, 2)
v = v.view(num_batches, -1, num_heads, d_k).transpose(1, 2)
print("q", q.size())
print("k", k.size())
print("v", v.size())
d_k = q.size(-1)
scores = torch.matmul(k, q.transpose(-2, -1)) / math.sqrt(d_k)
scores = scores.masked_fill(x_mask == 0, -epsilon) # note source mask
attn = F.softmax(scores, dim = -1)
y = torch.matmul(attn, v)
print(y.size())
```

组装头,代码如下:

```python
y = y.transpose(1, 2).contiguous().view(num_batches, -1, num_heads * (d_embed // num_heads))
print(y.size())
```

后源关注前馈,代码如下:

```python
ff_d2 = nn.Linear(d_embed, d_embed)
y = ff_d2(y)
print(y.size())
```

添加回残差,代码如下:

```python
y = y_residual + y
print(y.size())
```

留出残差,代码如下:

```python
y_residual = y.clone()
print(y.size())
```

预前馈层归一化,代码如下:

```python
mean = y.mean(-1, keepdim=True)
std = y.std(-1, keepdim=True)
W3_d = nn.Parameter(torch.ones(d_embed))
```

```
b3_d = nn.Parameter(torch.zeros(d_embed))
y = W3_d * (y - mean) / (std + epsilon) + b3_d
print(y.size())
```

前馈，代码如下：

```
linear_expand_d = nn.Linear(d_embed, d_ff)
linear_compress_d = nn.Linear(d_ff, d_embed)
y = linear_compress_d(F.relu(linear_expand_d(y)))
print(y.size())
```

添加回残差，代码如下：

```
y = y_residual + y
print(y.size())
```

最终解码器层规范化，代码如下：

```
mean = y.mean(-1, keepdim=True)
std = y.std(-1, keepdim=True)
Wn_d = nn.Parameter(torch.ones(d_embed))
bn_d = nn.Parameter(torch.zeros(d_embed))
y = Wn_d * (y - mean) / (std + epsilon) + bn_d
print(y.size())
```

3.4.3 生成概率分布

下一个模块位于解码器的顶部，并将解码器输出扩展为每个表征位置的词汇表上的对数概率分布。这对所有表征都是如此，尽管对损失计算唯一重要的是那些被屏蔽的表征。此处未进行损失计算，代码如下：

```
linear_scores = nn.Linear(d_embed, vocab)
probs = F.log_softmax(linear_scores(y), dim=-1)
print(probs.size())
```

3.5 为 PyTorch 模型提供服务

本节介绍如何将经过训练的 PyTorch 模型包装在 Flask 容器中，以便通过 Web API 将其公开。

Flask 是一个用 Python 编写的轻量级 Web 服务器。它为用户提供了一种方便的方法，可以快速设置 Web API，用于从经过训练的 PyTorch 模型中进行预测，既可以直接使用，也可以作为更大系统中的 Web 服务。

我们将创建一个 Web 服务，该服务接收图像，并将它们映射到 ImageNet 数据集的 1000 个类中的一个。要做到这一点，需要一个图像文件进行测试。还可以获得一个文件，该文件将把模型输出的类索引映射到人类可读的类名。

可以通过检查 TorchServe 存储库并将其复制到用户的工作文件夹来快速提取这两个支持文件。从控制台提示符中发出以下命令：

```
git clone https://github.com/pytorch/serve
copy .\serve\examples\image_classifier\kitten.jpg .\
copy .\serve\examples\image_classifier\index_to_name.json .\
```

导入相关包，代码如下：

```
# We'll be using a pre-trained DenseNet model from torchvision.models
import torchvision.models as models
# torchvision.transforms contains tools for manipulating your image data
import torchvision.transforms as transforms
# Pillow (PIL) is what we'll use to load the image file initially
from PIL import Image
# And of course we'll need classes from flask
from flask import Flask, jsonify, request
```

预处理，代码如下：

```
def transform_image(infile):
    input_transforms = [transforms.Resize(255),
        transforms.CenterCrop(224),
        transforms.ToTensor(),
        transforms.Normalize([0.485, 0.456, 0.406],
            [0.229, 0.224, 0.225])]
    my_transforms = transforms.Compose(input_transforms)
    image = Image.open(infile)
    timg = my_transforms(image)
    timg.unsqueeze_(0)
    return timg
```

Web 请求给了我们一个图像文件，但我们的模型期望 PyTorch 张量的形状是（N, 3, 224, 224），其中 N 是输入批次中的项目数。我们只有一个批大小为 1 的批。我们要做的第一件事是编写一组 TorchVision 变换来调整图像的大小并裁剪图像，将其转换为张量，然后归一化张量中的值。

之后，打开文件并应用变换。变换返回形状是 (3, 224, 224) 的张量，也就是 224×224 图像的 3 个颜色通道。因为需要使这张图像成为一个批，所以，使用 unsqueeze_(0) 调用来通过添加新的第一维来修改张量。张量现在具有形状（1, 3, 224, 224）。

通常，即使不使用图像数据，也需要将 HTTP 请求的输入转换为 PyTorch 可以使用的张量。

推理，代码如下：

```
def get_prediction(input_tensor):
    outputs = model.forward(input_tensor)
    _, y_hat = outputs.max(1)
    prediction = y_hat.item()
    return prediction
```

推理本身是最简单的部分：当将输入张量传递给图像分类模型时，会得到一个值的张量，这些值表示模型估计的图像属于特定类别的可能性。对 max() 的调用查找具有最大似然值的类，并返回该值和 ImageNet 类索引。最后，通过 item() 调用从包含类索引的张量中提取该类索引，并返回它。

后处理，代码如下：

```
def render_prediction(prediction_idx):
    stridx = str(prediction_idx)
    class_name = 'Unknown'
    if img_class_map is not None:
        if stridx in img_class_map is not None:
            class_name = img_class_map[stridx][1]

    return prediction_idx, class_name
```

render_prediction() 方法将预测的类索引映射到人类可读的类标签。

将以下内容粘贴到名为 app.py 的文件中：

```
import io
import json
import os

import torchvision.models as models
import torchvision.transforms as transforms
from PIL import Image
from flask import Flask, jsonify, request


app = Flask(__name__)
model = models.densenet121(pretrained=True)    # Trained on 1000 classes from ImageNet
model.eval()                                    # Turns off autograd



img_class_map = None
mapping_file_path = 'index_to_name.json'# Human-readable names for Imagenet classes
if os.path.isfile(mapping_file_path):
```

```python
        with open (mapping_file_path) as f:
            img_class_map = json.load(f)


    # Transform input into the form our model expects
    def transform_image(infile):
        input_transforms = [transforms.Resize(255),# We use multiple TorchVision transforms to ready the image
            transforms.CenterCrop(224),
            transforms.ToTensor(),
            transforms.Normalize([0.485, 0.456, 0.406],   # Standard normalization for ImageNet model input
                [0.229, 0.224, 0.225])]
        my_transforms = transforms.Compose(input_transforms)
        image = Image.open(infile)                        # Open the image file
        timg = my_transforms(image) # Transform PIL image to appropriately-shaped PyTorch tensor
        timg.unsqueeze_(0)    # PyTorch models expect batched input; create a batch of 1
        return timg


    def get_prediction(input_tensor):
        outputs = model.forward(input_tensor) # Get likelihoods for all ImageNet classes
        _, y_hat = outputs.max(1) # Extract the most likely class
        prediction = y_hat.item() # Extract the int value from the PyTorch tensor
        return prediction

    # Make the prediction human-readable
    def render_prediction(prediction_idx):
        stridx = str(prediction_idx)
        class_name = 'Unknown'
        if img_class_map is not None:
            if stridx in img_class_map is not None:
                class_name = img_class_map[stridx][1]

        return prediction_idx, class_name


    @app.route('/', methods=['GET'])
    def root():
        return jsonify({'msg' : 'Try POSTing to the /predict endpoint with an RGB image attachment'})
```

```python
@app.route('/predict', methods=['POST'])
def predict():
    if request.method == 'POST':
        file = request.files['file']
        if file is not None:
            input_tensor = transform_image(file)
            prediction_idx = get_prediction(input_tensor)
            class_id, class_name = render_prediction(prediction_idx)
            return jsonify({'class_id': class_id, 'class_name': class_name})

if __name__ == '__main__':
    app.run()
```

要想从控制台启动服务器，可使用以下命令：

```
set FLASK_APP= app.py
flask run
```

默认情况下，Flask 服务器正在侦听端口 5000。服务器运行后，打开另一个控制台窗口，测试新的推理服务器，代码如下：

```
curl -X POST -H "Content-Type: multipart/form-data" http://localhost:5000/predict -F "file=@kitten.jpg"
```

如果一切设置正确，应该收到类似以下内容的响应：

```
{"class_id":285,"class_name":"Egyptian_cat"}
```

3.6 本章小结

本章首先介绍了神经网络基础及卷积神经网络，然后介绍了 PyTorch 基础知识和 transformer 架构，最后介绍了如何为 PyTorch 模型提供服务。

1943 年，Warren McCulloch 和 Walter Pitts 创建了一种名为阈值逻辑的基于数学和算法的神经网络计算模型。1975 年，Werbos 的反向传播算法通过使多层网络的训练可行和有效，从而解决了 XOR 问题。反向传播是一种迭代算法，通过确定应该调整哪些权重和偏差来帮助最小化成本函数。

前馈神经网络是一种最简单的神经网络，各神经元分层排列。每个神经元只与前一层的神经元相连，接收前一层的输出，并输出给下一层。各层间没有反馈。

卷积神经网络的概念起源于 20 世纪 60 年代，当时加拿大神经科学家 David H. Hubel

和 Torsten Wiesel 通过对猫视觉皮层细胞的研究，提出了感受野的概念。这一发现为后续的神经网络设计提供了理论基础。1980 年前后，日本科学家福岛邦彦在 Hubel 和 Wiesel 工作的基础上，模拟生物视觉系统提出了一种层级化的多层人工神经网络，即"神经认知"（Neocognitron），被认为是现今卷积神经网络的前身。1989 年，杨立昆（Yann LeCun）提出了多层卷积神经网络 LeNet 用于手写数字识别，这是第一个真正意义上的卷积神经网络。LeNet 的提出标志着卷积神经网络开始应用于实际问题。随着计算机性能的提高和大数据技术的发展，卷积神经网络得到了广泛应用。2012 年，Alex Krizhevsky 等人提出的 AlexNet 在 ImageNet 比赛中取得了巨大的成功，引起了广泛关注。自此，卷积神经网络成为深度学习的重要组成部分，并取得了一系列突破性成果。自 AlexNet 之后，各种优秀的卷积神经网络结构相继被提出，如 VGG、Google Inception Net、ResNet 等，这些模型在图像识别等领域取得了良好的效果。

卷积神经网络是受语音信号处理中时延神经网络（TDNN）影响而发明的。

Python 编程语言最初并不是为数值计算而设计的，但早期引起了科学和工程界的关注，因此在 1995 年成立了一个名为 matrix-sig 的特殊兴趣小组，目的是定义一个矩阵计算包。一个矩阵包的实现由 Jim Fulton 完成，然后由 Jim Hugunin 推广成为 Numeric，也被称为 Numerical Python 扩展。

一个名为 Numarray 的新软件包为了成为 Numeric 的更灵活的替代品而编写出来。与 Numeric 一样，它现已被弃用。Numarray 对大型矩阵的运算速度更快，但在小型矩阵上比 Numeric 慢，因此，有一段时间这两个包用于不同的用例。最新版本的 Numeric v24.2 于 2005 年 11 月 11 日发布，Numarray v1.5.2 于 2006 年 8 月 24 日发布。

2005 年年初，NumPy 开发人员 Travis Oliphant 想要围绕一个矩阵包来统一社区，并将 Numarray 的功能移植到 Numeric，在 2006 年将结果发布为 NumPy 1.0。NumPy 这个新项目是 SciPy 的一部分。2011 年，NumPy 1.5.0 版本增加了对 Python 3 的支持。

第 4 章　PyTorch 开发深度学习应用

本章介绍文本分类、聊天机器人等自然语言处理应用。

4.1　文本分类

要使用 PyTorch 实现文本分类，可以执行以下步骤。

（1）预处理文本数据：包括将文本标记为单个单词或子单词，将其转换为数字表示，并创建词汇表。

（2）准备数据集：将数据集拆分为训练集、验证集和测试集。将文本数据转换为张量或数值表示，这些张量或数值表示可以输入模型中。

（3）设计模型架构：定义用于文本分类的神经网络架构。通常包括使用词嵌入来表示单词，然后是一个或多个层，如卷积层或递归层，最后是用于实现分类的完全连接层。

（4）训练模型：使用训练集来训练模型，包括正向传播、使用合适的损失函数（如交叉熵）计算损失，以及使用优化器（如 Adam 或 SGD）反向传播以更新模型的参数。

（5）评估模型：使用验证集来评估训练模型的性能。计算准确性、精确度、召回率或 F1 分数等指标，以评估模型的性能。

（6）测试模型：使用测试集评估模型在没见过的数据上的性能。这将使用户了解模型的泛化能力。

本节将设计一个简单的网络来对文本文档进行分类。将使用 torchtext 模块提供的 AG NEWS 数据集。除此之外，还将使用 torchtext 中提供的词汇构建和其他实用程序。本节的主要目的是介绍如何使用 PyTorch 和 torchtext 模块设计文本分类网络。

下面导入必要的库，并打印在本节中使用的版本，代码如下：

```
import torch

print("PyTorch Version : {}".format(torch.__version__))
```

输出结果如下：

```
PyTorch Version : 1.13.1+cpu
```

导入 torchtext 模块，代码如下：

```
import torchtext

print("Torch Text Version : {}".format(torchtext.__version__))
```

输出结果如下：

```
Torch Text Version : 0.14.1
```

4.1.1 准备数据集

本节逐步对文本数据集进行向量化，为神经网络做好准备。将文本数据转换为神经网络所需的实数向量列表。为了做到这一点，首先用数据表征填充词汇表，然后创建数据加载程序，每次调用时返回向量化数据。使用词频方法来向量化数据。

首先，加载 torchtext 提供的 AG NEWS 数据集。该数据集包含 4 个不同新闻类别的文本文档。数据集分为训练数据集和测试数据集。

```
from torch.utils.data import DataLoader

train_dataset, test_dataset = torchtext.datasets.AG_NEWS()

target_classes = ["World", "Sports", "Business", "Sci/Tec"]
```

然后，用来自训练数据集和测试数据集的数据表征（Token）填充词汇表。通过数据 get_tokenizer() 方法从 torchtext.data 模块获得第一个初始化的分词器。初始化一个简单的分词器，用于分隔单词和标点符号。

初始化分词器后，使用 torchtext.vocab 模块中提供的 build_vocab_from_iterator() 函数填充词汇表。该函数以迭代器作为输入，每次调用迭代器时，迭代器都会返回一个表征列表。我们创建了一个迭代器作为一个简单的函数，它以数据集列表作为输入。然后，它循环遍历每个数据集及其文本示例，产生使用分词器为每个示例生成的表征列表。还要求函数使用 <UNK> 表征作为特殊表征，词汇表中不存在的表征将被映射到该表征。

```
from torchtext.data import get_tokenizer
from torchtext.vocab import build_vocab_from_iterator

tokenizer = get_tokenizer("basic_english")

def build_vocab(datasets):
    for dataset in datasets:
        for _, text in dataset:
            yield tokenizer(text)
```

```
vocab = build_vocab_from_iterator(build_vocab([train_dataset, test_dataset]),
specials=["<UNK>"])
vocab.set_default_index(vocab["<UNK>"])
```

接下来，创建将在训练过程中使用的训练和测试数据加载程序。创建批处理大小为 256 的数据加载程序。两个数据加载程序都接收可调用的 collate_fn 参数。此函数负责对一批文本文档进行向量化。实现代码如下：

```
from sklearn.feature_extraction.text import TfidfVectorizer, CountVectorizer
from torch.utils.data import DataLoader
from torchtext.data.functional import to_map_style_dataset

vectorizer = CountVectorizer(vocabulary=vocab.get_itos(), tokenizer=tokenizer)

def vectorize_batch(batch):
    Y, X = list(zip(*batch))
    X = vectorizer.transform(X).todense()
    return torch.tensor(X, dtype=torch.float32), torch.tensor(Y) - 1

train_dataset, test_dataset  = torchtext.datasets.AG_NEWS()
train_dataset, test_dataset = to_map_style_dataset(train_dataset), to_map_style_dataset(test_dataset)

train_loader = DataLoader(train_dataset, batch_size=256, collate_fn=vectorize_batch)
test_loader = DataLoader(test_dataset, batch_size=256, collate_fn=vectorize_batch)
```

4.1.2 定义网络

本节使用 PyTorch 设计一个简单的线性层神经网络，使用它对文本文档进行分类。该网络将矢量化数据作为输入并返回预测。

该网络有 3 个线性层，分别具有 128、64 和 4 个输出单元。我们已经将 relu 激活，应用于前两个线性层的输出。该网络采用 PyTorch 的 Sequential API 进行设计。实现代码如下：

```
from torch import nn
from torch.nn import functional as F

class TextClassifier(nn.Module):
    def __init__(self):
        super(TextClassifier, self).__init__()
        self.seq = nn.Sequential(
            nn.Linear(len(vocab), 128),
            nn.ReLU(),
```

```
            nn.Linear(128, 64),
            nn.ReLU(),

            nn.Linear(64, 4),
            #nn.ReLU(),

            #nn.Linear(64, 4),
        )

    def forward(self, X_batch):
        return self.seq(X_batch)

text_classifier = TextClassifier()
for X, Y in train_loader:
    Y_preds = text_classifier(X)
    print(Y_preds.shape)
    break
```

4.1.3 训练网络

本节将对上一节中定义的网络进行训练。为了训练网络，设计了一个简单的函数，该函数将在调用时执行训练。该函数将模型、损失函数、优化器、训练数据加载程序、验证数据加载程序和回合数作为输入，然后执行训练循环若干次。对于每个回合，它使用训练数据加载器分批循环训练数据，该加载器为每个批次返回矢量化数据及其标签。对于每个批次，我们执行前向传递网络以进行预测、计算损失（使用预测和实际目标标签）、计算梯度并更新网络参数。该函数还记录每个批次的损失，并在每个回合结束时打印平均训练损失。我们还创建了另一个辅助函数，它以输入模型、损失函数和验证数据加载器来计算验证损失和准确性。

```
from tqdm import tqdm
from sklearn.metrics import accuracy_score
import gc

def CalcValLossAndAccuracy(model, loss_fn, val_loader):
    with torch.no_grad():
        Y_shuffled, Y_preds, losses = [],[],[]
        for X, Y in val_loader:
            preds = model(X)
            loss = loss_fn(preds, Y)
            losses.append(loss.item())
```

```python
            Y_shuffled.append(Y)
            Y_preds.append(preds.argmax(dim=-1))

        Y_shuffled = torch.cat(Y_shuffled)
        Y_preds = torch.cat(Y_preds)

        print("Valid Loss : {:.3f}".format(torch.tensor(losses).mean()))
        print("Valid Acc  : {:.3f}".format(accuracy_score(Y_shuffled.detach().numpy(), Y_preds.detach().numpy())))


def TrainModel(model, loss_fn, optimizer, train_loader, val_loader, epochs=10):
    for i in range(1, epochs+1):
        losses = []
        for X, Y in tqdm(train_loader):
            Y_preds = model(X)

            loss = loss_fn(Y_preds, Y)
            losses.append(loss.item())

            optimizer.zero_grad()
            loss.backward()
            optimizer.step()

        print("Train Loss : {:.3f}".format(torch.tensor(losses).mean()))
        CalcValLossAndAccuracy(model, loss_fn, val_loader)
```

将回合数初始化为 8，将学习率初始化为 0.0001。然后，初始化交叉熵损失、文本分类器网络和 Adam 优化器。最后，用必要的参数调用训练程序来进行训练。通过查看每个回合结束时的损失和准确性值，可以得出结论，我们的模型在文本分类任务中做得很好。可以执行各种超参数微调，以进一步提高网络的性能。

```python
from torch.optim import Adam

epochs = 8
learning_rate = 1e-4

loss_fn = nn.CrossEntropyLoss()
text_classifier = TextClassifier()
optimizer = Adam(text_classifier.parameters(), lr=learning_rate)

TrainModel(text_classifier, loss_fn, optimizer, train_loader, test_loader, epochs)
```

4.2 开发聊天机器人

本节将探索一个递归序列到序列模型的用例。将使用康奈尔电影对话语料库中的电影脚本来训练一个简单的聊天机器人。

会话模型是人工智能研究的热点。聊天机器人可以在各种设置中找到，包括客户服务应用程序和在线服务台。这些机器人通常由基于检索的模型驱动，该模型输出对特定形式问题的预定义响应。在像公司的 IT 服务台这样的高度受限的领域，这些模型可能已经足够了，但对于更通用的用例来说，它们还不够健壮。教机器在多个领域与人类进行有意义的对话是一个远未解决的研究问题。最近，深度学习热潮涌现了强大的生成模型，如谷歌的神经会话模型，这标志着向多领域生成会话模型迈出了一大步。本节将在 PyTorch 中实现这种模型。

在 https://zissou.infosci.cornell.edu/convokit/datasets/movie-corpus/movie-corpus.zip 下载数据 ZIP 文件。

```
# and put in a ``data/`` directory under the current directory.
#
# After that, let's import some necessities.
#

import torch
from torch.jit import script, trace
import torch.nn as nn
from torch import optim
import torch.nn.functional as F
import csv
import random
import re
import os
import unicodedata
import codecs
from io import open
import itertools
import math
import json


USE_CUDA = torch.cuda.is_available()
device = torch.device("cuda" if USE_CUDA else "cpu")
```

下一步是重新格式化数据文件，并将数据加载到可以使用的结构中。

康奈尔电影对话语料库是一个丰富的电影角色对话数据集：

- 10 292 对电影角色之间的 220 579 次对话交流。

- 617 部电影中的 9 035 个角色。
- 共 304 713 次发言。

这个数据集庞大而多样,语言形式、时间段、情感等都有很大的变化。我们希望这种多样性使模型对多种形式的输入和查询都具有稳健性。

首先,查看数据文件的一些行,以查看原始格式,代码如下:

```
corpus_name = "movie-corpus"
corpus = os.path.join("data", corpus_name)

def printLines(file, n=10):
    with open(file, 'rb') as datafile:
        lines = datafile.readlines()
    for line in lines[:n]:
        print(line)

printLines(os.path.join(corpus, "utterances.jsonl"))
```

输出结果如下:

```
b'{"id": "L1045", "conversation_id": "L1044", "text": "They do not!", "speaker": "u0", "meta": {"movie_id": "m0", "parsed": [{"rt": 1, "toks": [{"tok": "They", "tag": "PRP", "dep": "nsubj", "up": 1, "dn": []}, {"tok": "do", "tag": "VBP", "dep": "ROOT", "dn": [0, 2, 3]}, {"tok": "not", "tag": "RB", "dep": "neg", "up": 1, "dn": []}, {"tok": "!", "tag": ".", "dep": "punct", "up": 1, "dn": []}]}]}, "reply-to": "L1044", "timestamp": null, "vectors": []}\n'
b'{"id": "L1044", "conversation_id": "L1044", "text": "They do to!", "speaker": "u2", "meta": {"movie_id": "m0", "parsed": [{"rt": 1, "toks": [{"tok": "They", "tag": "PRP", "dep": "nsubj", "up": 1, "dn": []}, {"tok": "do", "tag": "VBP", "dep": "ROOT", "dn": [0, 2, 3]}, {"tok": "to", "tag": "TO", "dep": "dobj", "up": 1, "dn": []}, {"tok": "!", "tag": ".", "dep": "punct", "up": 1, "dn": []}]}]}, "reply-to": null, "timestamp": null, "vectors": []}\n'
b'{"id": "L985", "conversation_id": "L984", "text": "I hope so.", "speaker": "u0", "meta": {"movie_id": "m0", "parsed": [{"rt": 1, "toks": [{"tok": "I", "tag": "PRP", "dep": "nsubj", "up": 1, "dn": []}, {"tok": "hope", "tag": "VBP", "dep": "ROOT", "dn": [0, 2, 3]}, {"tok": "so", "tag": "RB", "dep": "advmod", "up": 1, "dn": []}, {"tok": ".", "tag": ".", "dep": "punct", "up": 1, "dn": []}]}]}, "reply-to": "L984", "timestamp": null, "vectors": []}\n'
b'{"id": "L984", "conversation_id": "L984", "text": "She okay?", "speaker": "u2", "meta": {"movie_id": "m0", "parsed": [{"rt": 1, "toks": [{"tok": "She", "tag": "PRP", "dep": "nsubj", "up": 1, "dn": []}, {"tok": "okay", "tag": "RB", "dep": "ROOT", "dn": [0, 2]}, {"tok": "?", "tag": ".", "dep": "punct", "up": 1, "dn": []}]}]}, "reply-to": null, "timestamp": null, "vectors": []}\n'
b'{"id": "L925", "conversation_id": "L924", "text": "Let\'s go.", "speaker": "u0", "meta": {"movie_id": "m0", "parsed": [{"rt": 0, "toks": [{"tok": "Let", "tag": "VB", "dep": "ROOT", "dn": [2, 3]}, {"tok": "\'s", "tag": "PRP", "dep": "nsubj",
```

```
"up": 2, "dn": []}, {"tok": "go", "tag": "VB", "dep": "ccomp", "up": 0, "dn": [1]},
{"tok": ".", "tag": ".", "dep": "punct", "up": 0, "dn": []}]}]}, "reply-to": "L924",
"timestamp": null, "vectors": []}\n'
    b'{"id": "L924", "conversation_id": "L924", "text": "Wow", "speaker": "u2",
"meta": {"movie_id": "m0", "parsed": [{"rt": 0, "toks": [{"tok": "Wow", "tag": "UH",
"dep": "ROOT", "dn": []}]}]}, "reply-to": null, "timestamp": null, "vectors": []}\n'
    b'{"id": "L872", "conversation_id": "L870", "text": "Okay -- you\'re gonna need
to learn how to lie.", "speaker": "u0", "meta": {"movie_id": "m0", "parsed": [{"rt":
4, "toks": [{"tok": "Okay", "tag": "UH", "dep": "intj", "up": 4, "dn": []}, {"tok":
"--", "tag": ":", "dep": "punct", "up": 4, "dn": []}, {"tok": "you", "tag": "PRP",
"dep": "nsubj", "up": 4, "dn": []}, {"tok": "\'re", "tag": "VBP", "dep": "aux", "up":
4, "dn": []}, {"tok": "gon", "tag": "VBG", "dep": "ROOT", "dn": [0, 1, 2, 3, 6, 12]},
{"tok": "na", "tag": "TO", "dep": "aux", "up": 6, "dn": []}, {"tok": "need", "tag":
"VB", "dep": "xcomp", "up": 4, "dn": [5, 8]}, {"tok": "to", "tag": "TO", "dep": "aux",
"up": 8, "dn": []}, {"tok": "learn", "tag": "VB", "dep": "xcomp", "up": 6, "dn": [7,
11]}, {"tok": "how", "tag": "WRB", "dep": "advmod", "up": 11, "dn": []}, {"tok":
"to", "tag": "TO", "dep": "aux", "up": 11, "dn": []}, {"tok": "lie", "tag": "VB",
"dep": "xcomp", "up": 8, "dn": [9, 10]}, {"tok": ".", "tag": ".", "dep": "punct",
"up": 4, "dn": []}]}]}, "reply-to": "L871", "timestamp": null, "vectors": []}\n'
    b'{"id": "L871", "conversation_id": "L870", "text": "No", "speaker": "u2",
"meta": {"movie_id": "m0", "parsed": [{"rt": 0, "toks": [{"tok": "No", "tag": "UH",
"dep": "ROOT", "dn": []}]}]}, "reply-to": "L870", "timestamp": null, "vectors": []}\n'
```

为了方便起见，将创建一个格式良好的数据文件，其中每一行都包含一个制表符分隔的查询语句和一对响应语句。

以下函数有助于解析原始 outternances.jsonl 数据文件。

- loadLinesAndConversations 函数将文件的每一行拆分为一个包含字段的行字典：lineID、characterID 和 text；然后将它们分组为包含字段的对话：conversationID、movieID 和行。
- extractSentencePairs 函数从对话中提取成对的句子。

```
# Splits each line of the file to create lines and conversations
def loadLinesAndConversations(fileName):
    lines = {}
    conversations = {}
    with open(fileName, 'r', encoding='iso-8859-1') as f:
        for line in f:
            lineJson = json.loads(line)
            # Extract fields for line object
            lineObj = {}
            lineObj["lineID"] = lineJson["id"]
            lineObj["characterID"] = lineJson["speaker"]
            lineObj["text"] = lineJson["text"]
            lines[lineObj['lineID']] = lineObj
```

```python
            # Extract fields for conversation object
            if lineJson["conversation_id"] not in conversations:
                convObj = {}
                convObj["conversationID"] = lineJson["conversation_id"]
                convObj["movieID"] = lineJson["meta"]["movie_id"]
                convObj["lines"] = [lineObj]
            else:
                convObj = conversations[lineJson["conversation_id"]]
                convObj["lines"].insert(0, lineObj)
            conversations[convObj["conversationID"]] = convObj

    return lines, conversations


# Extracts pairs of sentences from conversations
def extractSentencePairs(conversations):
    qa_pairs = []
    for conversation in conversations.values():
        # Iterate over all the lines of the conversation
        for i in range(len(conversation["lines"]) - 1):  # We ignore the last line (no answer for it)
            inputLine = conversation["lines"][i]["text"].strip()
            targetLine = conversation["lines"][i+1]["text"].strip()
            # Filter wrong samples (if one of the lists is empty)
            if inputLine and targetLine:
                qa_pairs.append([inputLine, targetLine])
    return qa_pairs
```

现在将调用这些函数并创建文件。我们将其称为 formatted_movie_lines.txt。

```python
# Define path to new file
datafile = os.path.join(corpus, "formatted_movie_lines.txt")

delimiter = '\t'
# Unescape the delimiter
delimiter = str(codecs.decode(delimiter, "unicode_escape"))

# Initialize lines dict and conversations dict
lines = {}
conversations = {}
# Load lines and conversations
print("\nProcessing corpus into lines and conversations...")
lines, conversations = loadLinesAndConversations(os.path.join(corpus, "utterances.jsonl"))
```

```
# Write new CSV file
print("\nWriting newly formatted file...")
with open(datafile, 'w', encoding='utf-8') as outputfile:
    writer = csv.writer(outputfile, delimiter=delimiter, lineterminator='\n')
    for pair in extractSentencePairs(conversations):
        writer.writerow(pair)

# Print a sample of lines
print("\nSample lines from file:")
printLines(datafile)
```

输出结果如下:

```
Processing corpus into lines and conversations...

Writing newly formatted file...

Sample lines from file:
b'They do to!\tThey do not!\n'
b'She okay?\tI hope so.\n'
b"Wow\tLet's go.\n"
b'"I\'m kidding.  You know how sometimes you just become this ""persona""?  And you don\'t know how to quit?"\tNo\n'
b"No\tOkay -- you're gonna need to learn how to lie.\n"
b'I figured you\'d get to the good stuff eventually.\tWhat good stuff?\n"
b'What good stuff?\t"The ""real you"".""\n'
b'"The ""real you""."\tLike my fear of wearing pastels?\n'
b'do you listen to this crap?\tWhat crap?\n'
b"What crap?\tMe.  This endless ...blonde babble. I'm like, boring myself.\n"
```

下一个任务是创建一个词汇表,并将查询/响应语句对加载到内存中。

我们处理的是单词序列,它们没有到离散数值空间的隐式映射,因此,必须通过将数据集中遇到的每个唯一单词映射到索引值来创建一个索引。

为此,定义了一个 Voc 类,它保持从单词到索引的映射、索引到单词的反向映射、每个单词的计数和总单词计数。Voc 类提供了向词汇表中添加单词(addWord)、添加句子中的所有单词(addSentence)和修剪不常见单词(trim)的方法。稍后将详细介绍修剪。

```
# Default word tokens
PAD_token = 0  # Used for padding short sentences
SOS_token = 1  # Start-of-sentence token
EOS_token = 2  # End-of-sentence token

class Voc:
    def __init__(self, name):
```

```python
        self.name = name
        self.trimmed = False
        self.word2index = {}
        self.word2count = {}
        self.index2word = {PAD_token: "PAD", SOS_token: "SOS", EOS_token: "EOS"}
        self.num_words = 3  # Count SOS, EOS, PAD

    def addSentence(self, sentence):
        for word in sentence.split(' '):
            self.addWord(word)

    def addWord(self, word):
        if word not in self.word2index:
            self.word2index[word] = self.num_words
            self.word2count[word] = 1
            self.index2word[self.num_words] = word
            self.num_words += 1
        else:
            self.word2count[word] += 1

    # Remove words below a certain count threshold
    def trim(self, min_count):
        if self.trimmed:
            return
        self.trimmed = True

        keep_words = []

        for k, v in self.word2count.items():
            if v >= min_count:
                keep_words.append(k)

        print('keep_words {} / {} = {:.4f}'.format(
            len(keep_words), len(self.word2index), len(keep_words) / len(self.word2index)
        ))

        # Reinitialize dictionaries
        self.word2index = {}
        self.word2count = {}
        self.index2word = {PAD_token: "PAD", SOS_token: "SOS", EOS_token: "EOS"}
        self.num_words = 3 # Count default tokens

        for word in keep_words:
            self.addWord(word)
```

现在可以组合词汇和查询/响应语句对。在准备使用这些数据之前，必须进行一些预处理。

首先，必须使用 unicodeToAscii() 函数将 Unicode 字符串转换为 ASCII；接下来，将所有字母转换为小写，并修剪除基本标点符号之外的所有非字母字符（normalizeString）；最后，为了有助于训练收敛性，将筛选出长度大于 MAX_LENGTH 阈值的句子（filterPairs）。

```
MAX_LENGTH = 10  # Maximum sentence length to consider

# Turn a Unicode string to plain ASCII, thanks to
# https://stackoverflow.com/a/518232/2809427
def unicodeToAscii(s):
    return ''.join(
        c for c in unicodedata.normalize('NFD', s)
        if unicodedata.category(c) != 'Mn'
    )

# Lowercase, trim, and remove non-letter characters
def normalizeString(s):
    s = unicodeToAscii(s.lower().strip())
    s = re.sub(r"([.!?])", r" \1", s)
    s = re.sub(r"[^a-zA-Z.!?]+", r" ", s)
    s = re.sub(r"\s+", r" ", s).strip()
    return s

# Read query/response pairs and return a voc object
def readVocs(datafile, corpus_name):
    print("Reading lines...")
    # Read the file and split into lines
    lines = open(datafile, encoding='utf-8').\
        read().strip().split('\n')
    # Split every line into pairs and normalize
    pairs = [[normalizeString(s) for s in l.split('\t')] for l in lines]
    voc = Voc(corpus_name)
    return voc, pairs

# Returns True if both sentences in a pair 'p' are under the MAX_LENGTH threshold
def filterPair(p):
    # Input sequences need to preserve the last word for EOS token
    return len(p[0].split(' ')) < MAX_LENGTH and len(p[1].split(' ')) < MAX_LENGTH

# Filter pairs using the ``filterPair`` condition
def filterPairs(pairs):
    return [pair for pair in pairs if filterPair(pair)]
```

```python
# Using the functions defined above, return a populated voc object and pairs list
def loadPrepareData(corpus, corpus_name, datafile, save_dir):
    print("Start preparing training data ...")
    voc, pairs = readVocs(datafile, corpus_name)
    print("Read {!s} sentence pairs".format(len(pairs)))
    pairs = filterPairs(pairs)
    print("Trimmed to {!s} sentence pairs".format(len(pairs)))
    print("Counting words...")
    for pair in pairs:
        voc.addSentence(pair[0])
        voc.addSentence(pair[1])
    print("Counted words:", voc.num_words)
    return voc, pairs


# Load/Assemble voc and pairs
save_dir = os.path.join("data", "save")
voc, pairs = loadPrepareData(corpus, corpus_name, datafile, save_dir)
# Print some pairs to validate
print("\npairs:")
for pair in pairs[:10]:
    print(pair)
```

输出结果如下:

```
Start preparing training data ...
Reading lines...
Read 221282 sentence pairs
Trimmed to 64313 sentence pairs
Counting words...
Counted words: 18082

pairs:
['they do to !', 'they do not !']
['she okay ?', 'i hope so .']
['wow', 'let s go .']
['what good stuff ?', 'the real you .']
['the real you .', 'like my fear of wearing pastels ?']
['do you listen to this crap ?', 'what crap ?']
['well no . . .', 'then that s all you had to say .']
['then that s all you had to say .', 'but']
['but', 'you always been this selfish ?']
['have fun tonight ?', 'tons']
```

另一种有利于在训练中更快收敛的策略是从词汇表中删除很少使用的单词。减少特征空间也将减轻模型必须学习近似的函数的难度。修剪单词将分如下两步进行：

（1）使用 voc.Trim 函数修剪在 MIN_COUNT 阈值以下使用的单词。

（2）筛选出带有修剪单词的配对。

```
MIN_COUNT = 3    # Minimum word count threshold for trimming

def trimRareWords(voc, pairs, MIN_COUNT):
    # Trim words used under the MIN_COUNT from the voc
    voc.trim(MIN_COUNT)
    # Filter out pairs with trimmed words
    keep_pairs = []
    for pair in pairs:
        input_sentence = pair[0]
        output_sentence = pair[1]
        keep_input = True
        keep_output = True
        # Check input sentence
        for word in input_sentence.split(' '):
            if word not in voc.word2index:
                keep_input = False
                break
        # Check output sentence
        for word in output_sentence.split(' '):
            if word not in voc.word2index:
                keep_output = False
                break

        # Only keep pairs that do not contain trimmed word(s) in their input or output sentence
        if keep_input and keep_output:
            keep_pairs.append(pair)

    print("Trimmed from {} pairs to {}, {:.4f} of total".format(len(pairs), len(keep_pairs), len(keep_pairs) / len(pairs)))
    return keep_pairs


# Trim voc and pairs
pairs = trimRareWords(voc, pairs, MIN_COUNT)
```

输出结果如下：

```
keep_words 7833 / 18079 = 0.4333
Trimmed from 64313 pairs to 53131, 0.8261 of total
```

尽管已经付出了大量的努力来准备数据，并将其整理成一个漂亮的词汇对象和句子对列表，但模型最终会期望数字 torch 张量作为输入。

如果简单地通过将单词转换为它们的索引（indexesFromSentence）和零填充来将英语句子转换为张量，那么张量将具有形状（batch_size，max_length），并且对第一维度进行索引将返回所有时间步长的完整序列。但是，需要能够沿时间对批次进行索引，并跨批次中的所有序列。因此，将输入的批处理形状转换为（max_length，batch_size），以便跨第一维度的索引返回跨批处理中所有句子的时间步长。在 zeroPadding() 函数中隐式处理这个转换。

inputVar() 函数处理将句子转换为张量的过程，最终创建一个形状正确的零填充张量。它还为批处理中的每个序列返回一个长度张量，稍后将传递给解码器。

outputVar() 函数执行与 inputVar 类似的功能，但它不是返回长度张量，而是返回二进制掩码张量和最大目标句子长度。二进制掩码张量与输出目标张量具有相同的形状，但 PAD_token 的每个元素都是 0，其他元素都是 1。

batch2TrainData() 函数只获取一组对，并使用上述函数返回输入张量和目标张量。

```python
def indexesFromSentence(voc, sentence):
    return [voc.word2index[word] for word in sentence.split(' ')] + [EOS_token]


def zeroPadding(l, fillvalue=PAD_token):
    return list(itertools.zip_longest(*l, fillvalue=fillvalue))


def binaryMatrix(l, value=PAD_token):
    m = []
    for i, seq in enumerate(l):
        m.append([])
        for token in seq:
            if token == PAD_token:
                m[i].append(0)
            else:
                m[i].append(1)
    return m


# Returns padded input sequence tensor and lengths
def inputVar(l, voc):
    indexes_batch = [indexesFromSentence(voc, sentence) for sentence in l]
    lengths = torch.tensor([len(indexes) for indexes in indexes_batch])
    padList = zeroPadding(indexes_batch)
    padVar = torch.LongTensor(padList)
    return padVar, lengths
```

```
# Returns padded target sequence tensor, padding mask, and max target length
def outputVar(l, voc):
    indexes_batch = [indexesFromSentence(voc, sentence) for sentence in l]
    max_target_len = max([len(indexes) for indexes in indexes_batch])
    padList = zeroPadding(indexes_batch)
    mask = binaryMatrix(padList)
    mask = torch.BoolTensor(mask)
    padVar = torch.LongTensor(padList)
    return padVar, mask, max_target_len

# Returns all items for a given batch of pairs
def batch2TrainData(voc, pair_batch):
    pair_batch.sort(key=lambda x: len(x[0].split(" ")), reverse=True)
    input_batch, output_batch = [], []
    for pair in pair_batch:
        input_batch.append(pair[0])
        output_batch.append(pair[1])
    inp, lengths = inputVar(input_batch, voc)
    output, mask, max_target_len = outputVar(output_batch, voc)
    return inp, lengths, output, mask, max_target_len


# Example for validation
small_batch_size = 5
batches = batch2TrainData(voc, [random.choice(pairs) for _ in range(small_batch_size)])
input_variable, lengths, target_variable, mask, max_target_len = batches

print("input_variable:", input_variable)
print("lengths:", lengths)
print("target_variable:", target_variable)
print("mask:", mask)
print("max_target_len:", max_target_len)
```

输出结果如下：

```
input_variable: tensor([[  86,   24,  140,  829,   62],
        [   6,  355, 1362,  206,  566],
        [  36,  735,   14,   72, 1919],
        [  17,  140,  140, 2160,   85],
        [  62,   28,  158,   14,   14],
        [1012,  461,  140,    2,    2],
        [3223,   10,   14,    0,    0],
        [1012,    2,    2,    0,    0],
        [   6,    0,    0,    0,    0],
        [   2,    0,    0,    0,    0]])
```

```
lengths: tensor([10, 8, 8, 6, 6])
target_variable: tensor([[  18,  11, 101,  93, 277],
        [ 483, 113,  19, 311,  72],
        [   5, 241,  10,  72,  10],
        [  22, 706,   2,  19,   2],
        [2010,  14,   0,  24,   0],
        [1556,   2,   0, 136,   0],
        [  14,   0,   0,   5,   0],
        [   2,   0,   0,  48,   0],
        [   0,   0,   0,  14,   0],
        [   0,   0,   0,   2,   0]])
mask: tensor([[ True,  True,  True,  True,  True],
        [ True,  True,  True,  True,  True],
        [ True,  True,  True,  True,  True],
        [ True,  True,  True,  True,  True],
        [ True,  True, False,  True, False],
        [ True,  True, False,  True, False],
        [ True, False, False,  True, False],
        [ True, False, False,  True, False],
        [False, False, False,  True, False],
        [False, False, False,  True, False]])
max_target_len: 10
```

聊天机器人的大脑是一个序列到序列（seq2seq）模型。seq2seq 模型的目标是将可变长度的序列作为输入，并使用固定大小的模型返回可变长度序列作为输出。

通过使用两个单独的递归神经网络，可以完成这项任务。一个 RNN 充当编码器，将可变长度的输入序列编码为固定长度的上下文向量。理论上，这个上下文向量（RNN 的最后隐藏层）将包含关于输入机器人的查询语句的语义信息。第二个 RNN 是解码器，它获取输入单词和上下文向量，并返回序列中下一个单词的猜测和隐藏状态，以便在下一次迭代中使用。

编码器 RNN 在输入句子中一次迭代一个表征（如单词），在每个时间步长输出一个输出向量和一个隐藏状态向量。然后将隐藏状态向量传递到下一个时间步长，同时记录输出向量。编码器将在序列中的每个点看到的上下文转换为高维空间中的一组点，解码器将使用这些点为给定任务生成有意义的输出。

编码器的核心是一个多层门控递归单元。这里将使用 GRU 的双向变体，这意味着本质上有两个独立的 RNN：一个按正常顺序输入序列，另一个按相反顺序输入序列。在每个时间步长对每个网络的输出求和。使用双向 GRU 将提供同时对过去和未来上下文进行编码的优势。

注意，嵌入层用于在任意大小的特征空间中对单词索引进行编码。对于模型，该层将把每个单词映射到大小为 hidden_size 的特征空间。当训练时，这些值应该对相似含义单词之间的语义相似性进行编码。

最后，如果将填充的一批序列传递给 RNN 模块，则必须分别使用 nn.utils.RNN.pack_padded_sequence 和 nn.utils.RNN.pad_packed_sequency 对 RNN 通道周围的填充进行打包和解包。

编辑器的计算步骤：

（1）将单词索引转换为嵌入。

（2）为 RNN 模块打包填充的序列批次。

（3）向前通过 GRU。

（4）打开填充。

（5）对双向 GRU 输出求和。

（6）返回输出和最终隐藏状态。

编辑器的输入如下：

- input_seq：输入语句的批次；批次的形状 shape=(max_length, batch_size)。
- input_lengths：批次中每个句子对应的句子长度列表；句子长度列表的形状 shape=(batch_size)。
- hidden：隐藏状态；隐藏状态的形状 shape=(n_layers × num_directions, batch_size, hidden_size)。

编辑器的输出如下：

- outputs：GRU 最后一个隐藏层的输出特征（双向输出的总和）输出特征的形状；shape=(max_length, batch_size, hidden_size)。
- hidden：从 GRU 更新隐藏状态；隐藏状态的形状 shape=(n_layers × num_directions, batch_size, hidden_size)。

```
class EncoderRNN(nn.Module):
    def __init__(self, hidden_size, embedding, n_layers=1, dropout=0):
        super(EncoderRNN, self).__init__()
        self.n_layers = n_layers
        self.hidden_size = hidden_size
        self.embedding = embedding

        # Initialize GRU; the input_size and hidden_size parameters are both set to 'hidden_size'
        #   because our input size is a word embedding with number of features == hidden_size
        self.gru = nn.GRU(hidden_size, hidden_size, n_layers,
                          dropout=(0 if n_layers == 1 else dropout), bidirectional=True)

    def forward(self, input_seq, input_lengths, hidden=None):
```

```python
# Convert word indexes to embeddings
embedded = self.embedding(input_seq)
# Pack padded batch of sequences for RNN module
packed = nn.utils.rnn.pack_padded_sequence(embedded, input_lengths)
# Forward pass through GRU
outputs, hidden = self.gru(packed, hidden)
# Unpack padding
outputs, _ = nn.utils.rnn.pad_packed_sequence(outputs)
# Sum bidirectional GRU outputs
outputs = outputs[:, :, :self.hidden_size] + outputs[:, : ,self.hidden_size:]
# Return output and final hidden state
return outputs, hidden
```

解码器 RNN 以逐个表征的方式生成响应语句。它使用编码器的上下文向量和内部隐藏状态来生成序列中的下一个单词。它继续生成单词，直到输出表示句子结尾的 EOS_token。原生 seq2seq 解码器的一个常见问题是，如果只依赖上下文向量来编码整个输入序列的含义，那么很可能会丢失信息。这在处理长输入序列时尤其如此，极大地限制了解码器的能力。

为了解决这一问题，Bahdanau 等人创建了一种注意力机制，允许解码器关注输入序列的某些部分，而不是在每一步都使用整个固定上下文。

在高级别上，使用解码器的当前隐藏状态和编码器的输出来计算注意力。输出注意力权重与输入序列具有相同的形状，允许将它们与编码器输出相乘，从而得到一个加权和，指示编码器输出中需要注意的部分。

Luong 等人通过创造"全局注意力"来改进 Bahdanau 等人的基础工作。关键的区别在于，对于"全局注意力"，考虑编码器的所有隐藏状态，而 Bahdanau 等人的"局部注意力"仅考虑编码器从当前时间步长的隐藏状态。另一个区别是，在"全局注意力"中，Luong 等人只使用当前时间步长的解码器隐藏状态来计算注意力权重或能量，而 Bahdanau 等人的注意力计算需要从上一个时间步长了解解码器的状态。

这里将把注意力层实现为一个名为 Attn 的独立 nn 模块。该模块的输出是具有形状（batch_size, 1, max_length）的 softmax 归一化权重张量。

```python
# Luong attention layer
class Attn(nn.Module):
    def __init__(self, method, hidden_size):
        super(Attn, self).__init__()
        self.method = method
        if self.method not in ['dot', 'general', 'concat']:
            raise ValueError(self.method, "is not an appropriate attention method.")
        self.hidden_size = hidden_size
        if self.method == 'general':
            self.attn = nn.Linear(self.hidden_size, hidden_size)
```

```python
        elif self.method == 'concat':
            self.attn = nn.Linear(self.hidden_size * 2, hidden_size)
            self.v = nn.Parameter(torch.FloatTensor(hidden_size))

    def dot_score(self, hidden, encoder_output):
        return torch.sum(hidden * encoder_output, dim=2)

    def general_score(self, hidden, encoder_output):
        energy = self.attn(encoder_output)
        return torch.sum(hidden * energy, dim=2)

    def concat_score(self, hidden, encoder_output):
        energy = self.attn(torch.cat((hidden.expand(encoder_output.size(0), -1, -1), encoder_output), 2)).tanh()
        return torch.sum(self.v * energy, dim=2)

    def forward(self, hidden, encoder_outputs):
        # Calculate the attention weights (energies) based on the given method
        if self.method == 'general':
            attn_energies = self.general_score(hidden, encoder_outputs)
        elif self.method == 'concat':
            attn_energies = self.concat_score(hidden, encoder_outputs)
        elif self.method == 'dot':
            attn_energies = self.dot_score(hidden, encoder_outputs)

        # Transpose max_length and batch_size dimensions
        attn_energies = attn_energies.t()

        # Return the softmax normalized probability scores (with added dimension)
        return F.softmax(attn_energies, dim=1).unsqueeze(1)
```

既然已经定义了注意力子模块，就可以实现实际的解码器模型了。对于解码器，将一次一个时间步长手动输入批次。这意味着我们嵌入的单词张量和 GRU 输出都将具有形状（1, batch_size, hidden_size）。

计算步骤：

（1）获取当前输入单词的嵌入。

（2）通过单向 GRU 向前馈送。

（3）根据步骤（2）中的当前 GRU 输出计算注意力权重。

（4）将注意力权重乘以编码器输出，得到新的"加权和"上下文向量。

（5）连接加权上下文向量和 GRU 输出。

（6）预测下一个单词。

（7）返回输出和最终隐藏状态。

输入：
- input_step：输入序列批的一个时间步长（一个词）；input_step 的形状 shape=(1, batch_size)。
- last_hidden：GRU 的最终隐藏层；last_hidden 的形状 shape=(n_layers × num_directions, batch_size, hidden_size)。
- encoder_outputs：编码器模型的输出；encoder_outputs 的形状 shape=(max_length, batch_size, hidden_size)。

输出：
- output：softmax 归一化张量，给出每个单词是解码序列中正确的下一个单词的概率；output 的形状 shape=(batch_size, voc.num_words)。
- hidden：GRU 的最终隐藏状态；hidden 的形状 shape=(n_layers × num_directions, batch_size, hidden_size)。

```python
class LuongAttnDecoderRNN(nn.Module):
    def __init__(self, attn_model, embedding, hidden_size, output_size, n_layers=1, dropout=0.1):
        super(LuongAttnDecoderRNN, self).__init__()

        # Keep for reference
        self.attn_model = attn_model
        self.hidden_size = hidden_size
        self.output_size = output_size
        self.n_layers = n_layers
        self.dropout = dropout

        # Define layers
        self.embedding = embedding
        self.embedding_dropout = nn.Dropout(dropout)
        self.gru = nn.GRU(hidden_size, hidden_size, n_layers, dropout=(0 if n_layers == 1 else dropout))
        self.concat = nn.Linear(hidden_size * 2, hidden_size)
        self.out = nn.Linear(hidden_size, output_size)

        self.attn = Attn(attn_model, hidden_size)

    def forward(self, input_step, last_hidden, encoder_outputs):
        # Note: we run this one step (word) at a time
        # Get embedding of current input word
        embedded = self.embedding(input_step)
        embedded = self.embedding_dropout(embedded)
        # Forward through unidirectional GRU
        rnn_output, hidden = self.gru(embedded, last_hidden)
```

```
            # Calculate attention weights from the current GRU output
            attn_weights = self.attn(rnn_output, encoder_outputs)
             # Multiply attention weights to encoder outputs to get new "weighted sum" context vector
            context = attn_weights.bmm(encoder_outputs.transpose(0, 1))
            # Concatenate weighted context vector and GRU output using Luong eq. 5
            rnn_output = rnn_output.squeeze(0)
            context = context.squeeze(1)
            concat_input = torch.cat((rnn_output, context), 1)
            concat_output = torch.tanh(self.concat(concat_input))
            # Predict next word using Luong eq. 6
            output = self.out(concat_output)
            output = F.softmax(output, dim=1)
            # Return output and final hidden state
            return output, hidden
```

由于处理的是填充序列，所以，在计算损失时，不能简单地考虑张量的所有元素。定义 maskNLLLoss() 函数，以基于解码器的输出张量、目标张量和描述目标张量填充的二进制掩码张量来计算损失。该损失函数计算对应于掩码张量中的 1 的元素的平均负对数似然。

```
    def maskNLLLoss(inp, target, mask):
        nTotal = mask.sum()
        crossEntropy = -torch.log(torch.gather(inp, 1, target.view(-1, 1)).squeeze(1))
        loss = crossEntropy.masked_select(mask).mean()
        loss = loss.to(device)
        return loss, nTotal.item()
```

训练函数包含单个训练迭代（单个输入批次）的算法。

可以使用几个巧妙的技巧来帮助收敛。

第一个技巧是使用教师强制。这意味着，在某种概率下，由 teacher_forging_ratio 设置，我们使用当前目标词作为解码器的下一个输入，而不是使用解码器的当前猜测。这种技术充当解码器的训练轮，有助于更有效的训练。然而，教师强迫可能会导致推理过程中的模型不稳定，因为解码器在训练过程中可能没有足够的机会真正制作自己的输出序列。因此，必须注意如何设置 teacher_forging_ratio，不要被快速收敛所欺骗。

第二个技巧是梯度剪裁。这是解决"爆炸梯度"问题的常用技术。本质上，通过将梯度剪裁或将梯度阈值设置为最大值，可以防止梯度呈指数级增长。

操作步骤如下：

（1）通过编码器向前传递整个输入批次。

（2）将解码器输入初始化为 SOS_token，将隐藏状态初始化为编码器的最终隐藏状态。

(3)通过解码器一次一个时间步长向前输入批处理序列。

(4)如果教师强制,将下一个解码器输入设置为当前目标;否则,将下一个解码器输入设置为当前解码器输出。

(5)计算并累计损失。

(6)执行反向传播。

(7)剪裁梯度。

(8)更新编码器和解码器模型参数。

```
def train(input_variable, lengths, target_variable, mask, max_target_len,
encoder, decoder, embedding,
          encoder_optimizer, decoder_optimizer, batch_size, clip, max_length=MAX_
LENGTH):

    # Zero gradients
    encoder_optimizer.zero_grad()
    decoder_optimizer.zero_grad()

    # Set device options
    input_variable = input_variable.to(device)
    target_variable = target_variable.to(device)
    mask = mask.to(device)
    # Lengths for RNN packing should always be on the CPU
    lengths = lengths.to("cpu")

    # Initialize variables
    loss = 0
    print_losses = []
    n_totals = 0

    # Forward pass through encoder
    encoder_outputs, encoder_hidden = encoder(input_variable, lengths)

    # Create initial decoder input (start with SOS tokens for each sentence)
    decoder_input = torch.LongTensor([[SOS_token for _ in range(batch_size)]])
    decoder_input = decoder_input.to(device)

    # Set initial decoder hidden state to the encoder's final hidden state
    decoder_hidden = encoder_hidden[:decoder.n_layers]

    # Determine if we are using teacher forcing this iteration
    use_teacher_forcing = True if random.random() < teacher_forcing_ratio else False

    # Forward batch of sequences through decoder one time step at a time
```

```python
        if use_teacher_forcing:
            for t in range(max_target_len):
                decoder_output, decoder_hidden = decoder(
                    decoder_input, decoder_hidden, encoder_outputs
                )
                # Teacher forcing: next input is current target
                decoder_input = target_variable[t].view(1, -1)
                # Calculate and accumulate loss
                mask_loss, nTotal = maskNLLLoss(decoder_output, target_variable[t], mask[t])
                loss += mask_loss
                print_losses.append(mask_loss.item() * nTotal)
                n_totals += nTotal
        else:
            for t in range(max_target_len):
                decoder_output, decoder_hidden = decoder(
                    decoder_input, decoder_hidden, encoder_outputs
                )
                # No teacher forcing: next input is decoder's own current output
                _, topi = decoder_output.topk(1)
                decoder_input = torch.LongTensor([[topi[i][0] for i in range(batch_size)]])
                decoder_input = decoder_input.to(device)
                # Calculate and accumulate loss
                mask_loss, nTotal = maskNLLLoss(decoder_output, target_variable[t], mask[t])
                loss += mask_loss
                print_losses.append(mask_loss.item() * nTotal)
                n_totals += nTotal

        # Perform backpropagation
        loss.backward()

        # Clip gradients: gradients are modified in place
        _ = nn.utils.clip_grad_norm_(encoder.parameters(), clip)
        _ = nn.utils.clip_grad_norm_(decoder.parameters(), clip)

        # Adjust model weights
        encoder_optimizer.step()
        decoder_optimizer.step()

        return sum(print_losses) / n_totals
```

下面将完整的训练过程与数据联系起来。trainIters() 函数负责在给定通过的模型、优化器、数据等的情况下运行训练的 n 次迭代。这个函数是不言自明的，因为我们已经用

train() 函数完成了繁重的工作。

需要注意的一点是，当保存模型时，会保存一个 tarball，其中包含编码器和解码器的 state_dicts（参数）、优化器的 state_dicts、损失、迭代等。以这种方式保存模型将提供检查点的终极灵活性。加载检查点后，将能够使用模型参数来运行推理，或者可以在停止的地方继续训练。

```python
def trainIters(model_name, voc, pairs, encoder, decoder, encoder_optimizer,
    decoder_optimizer, embedding, encoder_n_layers, decoder_n_layers, save_dir, n_
    iteration, batch_size, print_every, save_every, clip, corpus_name, loadFilename):

    # Load batches for each iteration
    training_batches = [batch2TrainData(voc, [random.choice(pairs) for _ in
range(batch_size)])
                        for _ in range(n_iteration)]

    # Initializations
    print('Initializing ...')
    start_iteration = 1
    print_loss = 0
    if loadFilename:
        start_iteration = checkpoint['iteration'] + 1

    # Training loop
    print("Training...")
    for iteration in range(start_iteration, n_iteration + 1):
        training_batch = training_batches[iteration - 1]
        # Extract fields from batch
        input_variable, lengths, target_variable, mask, max_target_len =
training_batch

        # Run a training iteration with batch
        loss = train(input_variable, lengths, target_variable, mask, max_target_
len, encoder,
                     decoder, embedding, encoder_optimizer, decoder_optimizer,
batch_size, clip)
        print_loss += loss

        # Print progress
        if iteration % print_every == 0:
            print_loss_avg = print_loss / print_every
            print("Iteration: {}; Percent complete: {:.1f}%; Average loss:
{:.4f}".format(iteration, iteration / n_iteration * 100, print_loss_avg))
            print_loss = 0
```

```
        # Save checkpoint
        if (iteration % save_every == 0):
            directory = os.path.join(save_dir, model_name, corpus_name, '{}-{}_
{}'.format(encoder_n_layers, decoder_n_layers, hidden_size))
            if not os.path.exists(directory):
                os.makedirs(directory)
            torch.save({
                'iteration': iteration,
                'en': encoder.state_dict(),
                'de': decoder.state_dict(),
                'en_opt': encoder_optimizer.state_dict(),
                'de_opt': decoder_optimizer.state_dict(),
                'loss': loss,
                'voc_dict': voc.__dict__,
                'embedding': embedding.state_dict()
             }, os.path.join(directory, '{}_{}.tar'.format(iteration,
'checkpoint')))
```

在训练了一个模型之后，希望能够自己与机器人对话。首先，必须定义希望模型如何解码编码的输入。

贪婪解码是在训练中不使用教师强迫时使用的解码方法。换句话说，对于每个时间步长，只需从 decoder_output 中选择 softmax 值最高的单词。这种解码方法在单个时间步长级别上是最优的。

为了便于贪婪解码操作，定义了一个 GreedySearchDecoder 类。当运行时，该类的对象采用形状是（input_seq length, 1）的输入序列（input_seq）、标量输入长度（input_length）张量和一个 max_length 来绑定响应语句长度。输入句子使用以下计算步骤进行评估。

（1）通过编码器模型进行正向输入。
（2）准备编码器的最终隐藏层作为解码器的第一个隐藏输入。
（3）将解码器的第一个输入初始化为 SOS_token。
（4）初始化张量以附加解码的单词。
（5）一次迭代解码一个单词表征。
①前向通过解码器。
②获取最可能的单词表征及其 softmax 分数。
③记录表征和分数。
④准备当前表征作为下一个解码器输入。
（6）返回单词表征和分数的集合。

```
class GreedySearchDecoder(nn.Module):
    def __init__(self, encoder, decoder):
        super(GreedySearchDecoder, self).__init__()
```

```python
        self.encoder = encoder
        self.decoder = decoder

    def forward(self, input_seq, input_length, max_length):
        # Forward input through encoder model
        encoder_outputs, encoder_hidden = self.encoder(input_seq, input_length)
        # Prepare encoder's final hidden layer to be first hidden input to the decoder
        decoder_hidden = encoder_hidden[:decoder.n_layers]
        # Initialize decoder input with SOS_token
        decoder_input = torch.ones(1, 1, device=device, dtype=torch.long) * SOS_token
        # Initialize tensors to append decoded words to
        all_tokens = torch.zeros([0], device=device, dtype=torch.long)
        all_scores = torch.zeros([0], device=device)
        # Iteratively decode one word token at a time
        for _ in range(max_length):
            # Forward pass through decoder
            decoder_output, decoder_hidden = self.decoder(decoder_input, decoder_hidden, encoder_outputs)
            # Obtain most likely word token and its softmax score
            decoder_scores, decoder_input = torch.max(decoder_output, dim=1)
            # Record token and score
            all_tokens = torch.cat((all_tokens, decoder_input), dim=0)
            all_scores = torch.cat((all_scores, decoder_scores), dim=0)
            # Prepare current token to be next decoder input (add a dimension)
            decoder_input = torch.unsqueeze(decoder_input, 0)
        # Return collections of word tokens and scores
        return all_tokens, all_scores
```

既然已经定义了解码方法，就可以编写用于评估字符串输入句子的函数。evaluate 函数管理着处理输入句子的底层过程。首先将句子格式化为 batch_size==1 的单词索引输入批。通过将句子中的单词转换为相应的索引，并转换维度为模型准备张量来实现这一点。还创建了一个长度张量，其中包含输入句子的长度。在这种情况下，长度是标量，因为一次只计算一个句子（batch_size==1）。接下来，使用 GreedySearchDecoder 对象（搜索器）获得解码的响应句子张量。最后，将响应的索引转换为单词，并返回解码单词的列表。

evaluateInput 充当聊天机器人的用户界面。调用时，将生成一个输入文本字段，可以在其中输入查询语句。在输入句子并按 Enter 键后，文本将以与训练数据相同的方式进行归一化，并最终输入评估函数以获得解码的输出句子。循环这个过程，就可以继续与机器人聊天，直到用户输入"q"或"quit"。

最后，如果输入的句子中不包含词汇表中的单词，则会通过打印错误消息并提示用户输入另一个句子来处理这一问题。

```python
def evaluate(encoder, decoder, searcher, voc, sentence, max_length=MAX_LENGTH):
```

```python
    ### Format input sentence as a batch
    # words -> indexes
    indexes_batch = [indexesFromSentence(voc, sentence)]
    # Create lengths tensor
    lengths = torch.tensor([len(indexes) for indexes in indexes_batch])
    # Transpose dimensions of batch to match models' expectations
    input_batch = torch.LongTensor(indexes_batch).transpose(0, 1)
    # Use appropriate device
    input_batch = input_batch.to(device)
    lengths = lengths.to("cpu")
    # Decode sentence with searcher
    tokens, scores = searcher(input_batch, lengths, max_length)
    # indexes -> words
    decoded_words = [voc.index2word[token.item()] for token in tokens]
    return decoded_words


def evaluateInput(encoder, decoder, searcher, voc):
    input_sentence = ''
    while(1):
        try:
            # Get input sentence
            input_sentence = input('> ')
            # Check if it is quit case
            if input_sentence == 'q' or input_sentence == 'quit': break
            # Normalize sentence
            input_sentence = normalizeString(input_sentence)
            # Evaluate sentence
            output_words = evaluate(encoder, decoder, searcher, voc, input_sentence)
            # Format and print response sentence
            output_words[:] = [x for x in output_words if not (x == 'EOS' or x == 'PAD')]
            print('Bot:', ' '.join(output_words))

        except KeyError:
            print("Error: Encountered unknown word.")
```

不管是想训练还是测试聊天机器人模型，都必须初始化各个编码器和解码器模型。在下面的代码块中设置所需的配置，选择从头开始或设置要加载的检查点，并构建和初始化模型。可以随意使用不同的模型配置来优化性能。

```python
# Configure models
model_name = 'cb_model'
attn_model = 'dot'
#``attn_model = 'general'``
#``attn_model = 'concat'``
```

```
hidden_size = 500
encoder_n_layers = 2
decoder_n_layers = 2
dropout = 0.1
batch_size = 64

# Set checkpoint to load from; set to None if starting from scratch
loadFilename = None
checkpoint_iter = 4000
```

从检查点加载的示例代码:

```
loadFilename = os.path.join(save_dir, model_name, corpus_name,
                '{}-{}_{}'.format(encoder_n_layers, decoder_n_layers, hidden_size),
                '{}_checkpoint.tar'.format(checkpoint_iter))

# Load model if a ``loadFilename`` is provided
if loadFilename:
    # If loading on same machine the model was trained on
    checkpoint = torch.load(loadFilename)
    # If loading a model trained on GPU to CPU
    #checkpoint = torch.load(loadFilename, map_location=torch.device('cpu'))
    encoder_sd = checkpoint['en']
    decoder_sd = checkpoint['de']
    encoder_optimizer_sd = checkpoint['en_opt']
    decoder_optimizer_sd = checkpoint['de_opt']
    embedding_sd = checkpoint['embedding']
    voc.__dict__ = checkpoint['voc_dict']


print('Building encoder and decoder ...')
# Initialize word embeddings
embedding = nn.Embedding(voc.num_words, hidden_size)
if loadFilename:
    embedding.load_state_dict(embedding_sd)
# Initialize encoder & decoder models
encoder = EncoderRNN(hidden_size, embedding, encoder_n_layers, dropout)
decoder = LuongAttnDecoderRNN(attn_model, embedding, hidden_size, voc.num_words, decoder_n_layers, dropout)
if loadFilename:
    encoder.load_state_dict(encoder_sd)
    decoder.load_state_dict(decoder_sd)
# Use appropriate device
encoder = encoder.to(device)
```

```
decoder = decoder.to(device)
print('Models built and ready to go!')
```

输出结果如下:

```
Building encoder and decoder ...
Models built and ready to go!
```

如果要训练模型,可运行以下代码块。

首先设置训练参数,然后初始化优化器,最后调用 trainIters 函数来运行训练迭代。

```
# Configure training/optimization
clip = 50.0
teacher_forcing_ratio = 1.0
learning_rate = 0.0001
decoder_learning_ratio = 5.0
n_iteration = 4000
print_every = 1
save_every = 500

# Ensure dropout layers are in train mode
encoder.train()
decoder.train()

# Initialize optimizers
print('Building optimizers ...')
encoder_optimizer = optim.Adam(encoder.parameters(), lr=learning_rate)
decoder_optimizer = optim.Adam(decoder.parameters(), lr=learning_rate * decoder_learning_ratio)
if loadFilename:
    encoder_optimizer.load_state_dict(encoder_optimizer_sd)
    decoder_optimizer.load_state_dict(decoder_optimizer_sd)

# If you have CUDA, configure CUDA to call
for state in encoder_optimizer.state.values():
    for k, v in state.items():
        if isinstance(v, torch.Tensor):
            state[k] = v.cuda()

for state in decoder_optimizer.state.values():
    for k, v in state.items():
        if isinstance(v, torch.Tensor):
            state[k] = v.cuda()

# Run training iterations
print("Starting Training!")
```

```
trainIters(model_name, voc, pairs, encoder, decoder, encoder_optimizer, decoder_
optimizer,
           embedding, encoder_n_layers, decoder_n_layers, save_dir, n_iteration,
batch_size,
           print_every, save_every, clip, corpus_name, loadFilename)
```

要想与模型聊天,可运行以下代码块。

```
# Set dropout layers to ``eval`` mode
encoder.eval()
decoder.eval()

# Initialize search module
searcher = GreedySearchDecoder(encoder, decoder)

# Begin chatting (uncomment and run the following line to begin)
# evaluateInput(encoder, decoder, searcher, voc)
```

4.3 用 Wav2Vec 2.0 进行语音识别

本节介绍如何使用 Wav2Vec 2.0 中的预训练模型进行语音识别。

语音识别的过程如下:

(1)从音频波形中提取声学特征。

(2)逐帧估计声学特征的类别。

(3)根据类概率序列生成假设。

Torchaudio 提供了对预先训练的权重和相关信息的轻松访问。它们捆绑在一起,可在 torchaudio.pipelines 模块下使用。

准备如下:

```
import torch
import torchaudio

print(torch.__version__)
print(torchaudio.__version__)

torch.random.manual_seed(0)
device = torch.device("cuda" if torch.cuda.is_available() else "cpu")

print(device)
```

输出如下:

```
2.0.0
2.0.1
cpu
```

下载资源:

```
import IPython
import matplotlib.pyplot as plt
from torchaudio.utils import download_asset

SPEECH_FILE = download_asset("tutorial-assets/Lab41-SRI-VOiCES-src-sp0307-ch127535-sg0042.wav")
```

输出如下:

```
  0%|          | 0.00/106k [00:00<?, ?B/s]
100%|##########| 106k/106k [00:00<00:00, 54.7MB/s]
```

首先,创建一个执行特征提取和分类的 Wav2Vec2 模型。

torchaudio 有两种类型的 Wav2Vec2 预训练权重。一些针对 ASR 任务进行了微调,另一些没有进行微调。

Wav2Vec2 模型以自监督的方式进行训练。它们首先使用音频进行训练,仅用于表示学习;然后使用附加标签对特定任务进行微调。

没有微调的预先训练的权重也可以针对其他下游任务进行微调,但本节并不涵盖这一点。将在此处使用 torchaudio.pipelines.WAV2VEC2_ASR_BAS_960H。torchaudio.pipelines 中提供了多个预训练模型。bundle 对象提供了实例化模型和其他信息的接口。采样率和类别标签如下:

```
bundle = torchaudio.pipelines.WAV2VEC2_ASR_BASE_960H

print("Sample Rate:", bundle.sample_rate)

print("Labels:", bundle.get_labels())
```

输出结果如下:

```
Sample Rate: 16000
Labels: ('-', '|', 'E', 'T', 'A', 'O', 'N', 'I', 'H', 'S', 'R', 'D', 'L', 'U', 'M', 'W', 'C', 'F', 'G', 'Y', 'P', 'B', 'V', 'K', "'", 'X', 'J', 'Q', 'Z')
```

模型可以构造如下:

```
model = bundle.get_model().to(device)

print(model.__class__)
```

此过程将自动获取预先训练的权重并将其加载到模型中。

输出结果如下:

```
Downloading: "https://download.pytorch.org/torchaudio/models/wav2vec2_fairseq_
base_ls960_asr_ls960.pth" to /root/.cache/torch/hub/checkpoints/wav2vec2_fairseq_
base_ls960_asr_ls960.pth

  0%|            | 0.00/360M [00:00<?, ?B/s]
 10%|9           | 36.0M/360M [00:00<00:00, 378MB/s]
 21%|##1         | 76.4M/360M [00:00<00:00, 404MB/s]
 33%|###2        | 119M/360M [00:00<00:00, 423MB/s]
 44%|####4       | 160M/360M [00:00<00:00, 425MB/s]
 56%|#####5      | 200M/360M [00:00<00:00, 414MB/s]
 67%|######6     | 240M/360M [00:00<00:00, 415MB/s]
 78%|#######7    | 279M/360M [00:00<00:00, 412MB/s]
 88%|########8   | 319M/360M [00:00<00:00, 402MB/s]
 99%|#########9  | 358M/360M [00:00<00:00, 406MB/s]
100%|##########| 360M/360M [00:00<00:00, 409MB/s]
<class 'torchaudio.models.wav2vec2.model.Wav2Vec2Model'>
```

将使用 VOiCES 数据集的语音数据,该数据集是根据 Creative Commos BY 4.0 授权的。

```
IPython.display.Audio(SPEECH_FILE)
```

要加载数据,可使用 torchaudio.load()。

如果采样率与管道所期望的不同,则可以使用 torchaudio.functional.resample() 进行重新采样。

```
waveform, sample_rate = torchaudio.load(SPEECH_FILE)
waveform = waveform.to(device)

if sample_rate != bundle.sample_rate:
    waveform = torchaudio.functional.resample(waveform, sample_rate, bundle.sample_rate)
```

下一步是从音频中提取声学特征。

```
with torch.inference_mode():
    features, _ = model.extract_features(waveform)
```

返回的特征是张量列表。每个张量都是 transformer 层的输出。

```
fig, ax = plt.subplots(len(features), 1, figsize=(16, 4.3 * len(features)))
for i, feats in enumerate(features):
    ax[i].imshow(feats[0].cpu(), interpolation="nearest")
    ax[i].set_title(f"Feature from transformer layer {i+1}")
    ax[i].set_xlabel("Feature dimension")
    ax[i].set_ylabel("Frame (time-axis)")
```

```
plt.tight_layout()
plt.show()
```

提取了声学特征后,下一步就是将它们分类到一组类别中。
Wav2Vec2 模型提供了一种一步完成特征提取和分类的方法。

```
with torch.inference_mode():
    emission, _ = model(waveform)
```

输出的形式是 logits。它不是以概率的形式出现的。

根据标记概率的序列,现在想要生成转录本(transcript)。产生假设的过程通常被称为解码。

解码比简单的分类更精细,因为在特定的时间步长解码可能会受到周围观测的影响。

例如,night 和 knight。即使它们的先验概率分布不同(在典型的对话中,night 的发生频率会比 knight 高),为了准确地生成 knight 的转录本,如 a knight with a sword,解码过程必须推迟最终决定,直到它看到足够的上下文。

存在许多解码技术,它们需要外部资源,如词典和语言模型。

为了简单起见,将执行不依赖于此类外部组件的贪婪解码,并在每个时间步长简单地选取最佳假设。因此,不使用上下文信息,并且只能生成一个转录本。

从定义贪婪解码算法开始。

```
class GreedyCTCDecoder(torch.nn.Module):
    def __init__(self, labels, blank=0):
        super().__init__()
        self.labels = labels
        self.blank = blank

    def forward(self, emission: torch.Tensor) -> str:
        """Given a sequence emission over labels, get the best path string
        Args:
          emission (Tensor): Logit tensors. Shape `[num_seq, num_label]`.

        Returns:
          str: The resulting transcript
        """
        indices = torch.argmax(emission, dim=-1)  # [num_seq,]
        indices = torch.unique_consecutive(indices, dim=-1)
        indices = [i for i in indices if i != self.blank]
        return "".join([self.labels[i] for i in indices])
```

现在创建解码器对象并解码转录本。

```
decoder = GreedyCTCDecoder(labels=bundle.get_labels())
transcript = decoder(emission[0])
```

检查一下结果,再听一遍音频。

```
print(transcript)
IPython.display.Audio(SPEECH_FILE)
```

ASR 模型使用名为 Connectionist Temporal Classification（CTC）的损失函数进行微调。在 CTC 中,空白标记（ϵ）是一种特殊的标记,表示前一个符号的重复。在解码过程中,这些会被忽略。

4.4 机器翻译

本节介绍如何使用 transformer 从头开始训练翻译模型。

torchtext 库具有用于创建数据集的实用程序,这些数据集可以很容易地迭代以创建语言翻译模型。本例展示了如何使用 torchtext 的内置数据集,标记原始文本句子,构建词汇表,并将表征数字化为张量。将使用 torchtext 库中的 Multi30k 数据集,该数据集生成一对源-目标原始句子。

要访问 torchtext 数据集,请安装 torchdata,代码如下:

```
from torchtext.data.utils import get_tokenizer
from torchtext.vocab import build_vocab_from_iterator
from torchtext.datasets import multi30k, Multi30k
from typing import Iterable, List


# We need to modify the URLs for the dataset since the links to the original dataset are broken
# Refer to https://github.com/pytorch/text/issues/1756#issuecomment-1163664163
# for more info
multi30k.URL["train"] = "https://raw.githubusercontent.com/neychev/small_DL_repo/master/datasets/Multi30k/training.tar.gz"
multi30k.URL["valid"] = "https://raw.githubusercontent.com/neychev/small_DL_repo/master/datasets/Multi30k/validation.tar.gz"

SRC_LANGUAGE = 'de'
TGT_LANGUAGE = 'en'

# Place-holders
token_transform = {}
vocab_transform = {}
```

创建源语言和目标语言标记器（请确保安装依赖项）,代码如下:

```
pip install -U torchdata
pip install -U spacy
python -m spacy download en_core_web_sm
python -m spacy download de_core_news_sm
```

transformer 是 *Attention is all you need* 论文中介绍的 Seq2Seq 模型，用于解决机器翻译任务。下面，将创建一个使用 transformer 的 Seq2Seq 网络。该网络由三部分组成。第一部分是嵌入层。该层将输入索引的张量转换为输入嵌入的相应张量。利用位置编码进一步增强这些嵌入，以向模型提供输入表征的位置信息。第二部分是实际的 transformer 模型。最后，transformer 模型的输出通过线性层，线性层为目标语言中的每个表征提供未规范化的概率。

```
from torch import Tensor
import torch
import torch.nn as nn
from torch.nn import Transformer
import math
DEVICE = torch.device('cuda' if torch.cuda.is_available() else 'cpu')


# helper Module that adds positional encoding to the token embedding to introduce a notion of word order.
class PositionalEncoding(nn.Module):
    def __init__(self,
                 emb_size: int,
                 dropout: float,
                 maxlen: int = 5000):
        super(PositionalEncoding, self).__init__()
        den = torch.exp(- torch.arange(0, emb_size, 2)* math.log(10000) / emb_size)
        pos = torch.arange(0, maxlen).reshape(maxlen, 1)
        pos_embedding = torch.zeros((maxlen, emb_size))
        pos_embedding[:, 0::2] = torch.sin(pos * den)
        pos_embedding[:, 1::2] = torch.cos(pos * den)
        pos_embedding = pos_embedding.unsqueeze(-2)

        self.dropout = nn.Dropout(dropout)
        self.register_buffer('pos_embedding', pos_embedding)

    def forward(self, token_embedding: Tensor):
        return self.dropout(token_embedding + self.pos_embedding[:token_embedding.size(0), :])

# helper Module to convert tensor of input indices into corresponding tensor of token embeddings
class TokenEmbedding(nn.Module):
```

```python
    def __init__(self, vocab_size: int, emb_size):
        super(TokenEmbedding, self).__init__()
        self.embedding = nn.Embedding(vocab_size, emb_size)
        self.emb_size = emb_size

    def forward(self, tokens: Tensor):
        return self.embedding(tokens.long()) * math.sqrt(self.emb_size)

# Seq2Seq Network
class Seq2SeqTransformer(nn.Module):
    def __init__(self,
                 num_encoder_layers: int,
                 num_decoder_layers: int,
                 emb_size: int,
                 nhead: int,
                 src_vocab_size: int,
                 tgt_vocab_size: int,
                 dim_feedforward: int = 512,
                 dropout: float = 0.1):
        super(Seq2SeqTransformer, self).__init__()
        self.transformer = Transformer(d_model=emb_size,
                                       nhead=nhead,
                                       num_encoder_layers=num_encoder_layers,
                                       num_decoder_layers=num_decoder_layers,
                                       dim_feedforward=dim_feedforward,
                                       dropout=dropout)
        self.generator = nn.Linear(emb_size, tgt_vocab_size)
        self.src_tok_emb = TokenEmbedding(src_vocab_size, emb_size)
        self.tgt_tok_emb = TokenEmbedding(tgt_vocab_size, emb_size)
        self.positional_encoding = PositionalEncoding(
            emb_size, dropout=dropout)

    def forward(self,
                src: Tensor,
                trg: Tensor,
                src_mask: Tensor,
                tgt_mask: Tensor,
                src_padding_mask: Tensor,
                tgt_padding_mask: Tensor,
                memory_key_padding_mask: Tensor):
        src_emb = self.positional_encoding(self.src_tok_emb(src))
        tgt_emb = self.positional_encoding(self.tgt_tok_emb(trg))
        outs = self.transformer(src_emb, tgt_emb, src_mask, tgt_mask, None,
                                src_padding_mask, tgt_padding_mask, memory_key_padding_mask)
```

```
        return self.generator(outs)

    def encode(self, src: Tensor, src_mask: Tensor):
        return self.transformer.encoder(self.positional_encoding(
                            self.src_tok_emb(src)), src_mask)

    def decode(self, tgt: Tensor, memory: Tensor, tgt_mask: Tensor):
        return self.transformer.decoder(self.positional_encoding(
                            self.tgt_tok_emb(tgt)), memory,
                            tgt_mask)
```

在训练过程中,需要一个后续的单词掩码,以防止模型在进行预测时查看未来的单词。还需要掩码来隐藏源和目标填充表征。下面定义一个同时处理这两个问题的函数,代码如下:

```
def generate_square_subsequent_mask(sz):
    mask = (torch.triu(torch.ones((sz, sz), device=DEVICE)) == 1).transpose(0, 1)
    mask = mask.float().masked_fill(mask == 0, float('-inf')).masked_fill(mask == 1, float(0.0))
    return mask


def create_mask(src, tgt):
    src_seq_len = src.shape[0]
    tgt_seq_len = tgt.shape[0]

    tgt_mask = generate_square_subsequent_mask(tgt_seq_len)
    src_mask = torch.zeros((src_seq_len, src_seq_len),device=DEVICE).type(torch.bool)

    src_padding_mask = (src == PAD_IDX).transpose(0, 1)
    tgt_padding_mask = (tgt == PAD_IDX).transpose(0, 1)
    return src_mask, tgt_mask, src_padding_mask, tgt_padding_mask
```

现在定义模型的参数并实例化它。还定义了损失函数,即交叉熵损失和用于训练的优化器。

```
torch.manual_seed(0)

SRC_VOCAB_SIZE = len(vocab_transform[SRC_LANGUAGE])
TGT_VOCAB_SIZE = len(vocab_transform[TGT_LANGUAGE])
EMB_SIZE = 512
NHEAD = 8
FFN_HID_DIM = 512
BATCH_SIZE = 128
```

```python
NUM_ENCODER_LAYERS = 3
NUM_DECODER_LAYERS = 3

transformer = Seq2SeqTransformer(NUM_ENCODER_LAYERS, NUM_DECODER_LAYERS, EMB_SIZE,
                                 NHEAD, SRC_VOCAB_SIZE, TGT_VOCAB_SIZE, FFN_HID_DIM)

for p in transformer.parameters():
    if p.dim() > 1:
        nn.init.xavier_uniform_(p)

transformer = transformer.to(DEVICE)

loss_fn = torch.nn.CrossEntropyLoss(ignore_index=PAD_IDX)

optimizer = torch.optim.Adam(transformer.parameters(), lr=0.0001, betas=(0.9, 0.98), eps=1e-9)
```

数据迭代器生成一对原始字符串。需要将这些字符串对转换为可以由之前定义的 Seq2Seq 网络处理的批处理张量。下面定义 collate() 函数，该函数将一批原始字符串转换为可以直接输入模型中的批张量。

```python
from torch.nn.utils.rnn import pad_sequence

# helper function to club together sequential operations
def sequential_transforms(*transforms):
    def func(txt_input):
        for transform in transforms:
            txt_input = transform(txt_input)
        return txt_input
    return func

# function to add BOS/EOS and create tensor for input sequence indices
def tensor_transform(token_ids: List[int]):
    return torch.cat((torch.tensor([BOS_IDX]),
                      torch.tensor(token_ids),
                      torch.tensor([EOS_IDX])))

# ``src`` and ``tgt`` language text transforms to convert raw strings into tensors indices
text_transform = {}
for ln in [SRC_LANGUAGE, TGT_LANGUAGE]:
    text_transform[ln] = sequential_transforms(token_transform[ln], #Tokenization
                                               vocab_transform[ln], #Numericalization
```

```
                                              tensor_transform) # Add BOS/EOS
and create tensor


    # function to collate data samples into batch tensors
    def collate_fn(batch):
        src_batch, tgt_batch = [], []
        for src_sample, tgt_sample in batch:
            src_batch.append(text_transform[SRC_LANGUAGE](src_sample.rstrip("\n")))
            tgt_batch.append(text_transform[TGT_LANGUAGE](tgt_sample.rstrip("\n")))

        src_batch = pad_sequence(src_batch, padding_value=PAD_IDX)
        tgt_batch = pad_sequence(tgt_batch, padding_value=PAD_IDX)
        return src_batch, tgt_batch
```

定义将为每个回合调用的训练和评估循环。

```
    from torch.utils.data import DataLoader

    def train_epoch(model, optimizer):
        model.train()
        losses = 0
        train_iter = Multi30k(split='train', language_pair=(SRC_LANGUAGE, TGT_LANGUAGE))
        train_dataloader = DataLoader(train_iter, batch_size=BATCH_SIZE, collate_fn=collate_fn)

        for src, tgt in train_dataloader:
            src = src.to(DEVICE)
            tgt = tgt.to(DEVICE)

            tgt_input = tgt[:-1, :]

            src_mask, tgt_mask, src_padding_mask, tgt_padding_mask = create_mask(src, tgt_input)

            logits = model(src, tgt_input, src_mask, tgt_mask,src_padding_mask, tgt_padding_mask, src_padding_mask)

            optimizer.zero_grad()

            tgt_out = tgt[1:, :]
            loss = loss_fn(logits.reshape(-1, logits.shape[-1]), tgt_out.reshape(-1))
            loss.backward()

            optimizer.step()
```

```
            losses += loss.item()

    return losses / len(list(train_dataloader))


def evaluate(model):
    model.eval()
    losses = 0

    val_iter = Multi30k(split='valid', language_pair=(SRC_LANGUAGE, TGT_LANGUAGE))
    val_dataloader = DataLoader(val_iter, batch_size=BATCH_SIZE, collate_fn=collate_fn)

    for src, tgt in val_dataloader:
        src = src.to(DEVICE)
        tgt = tgt.to(DEVICE)

        tgt_input = tgt[:-1, :]

        src_mask, tgt_mask, src_padding_mask, tgt_padding_mask = create_mask(src, tgt_input)

        logits = model(src, tgt_input, src_mask, tgt_mask, src_padding_mask, tgt_padding_mask, src_padding_mask)

        tgt_out = tgt[1:, :]
        loss = loss_fn(logits.reshape(-1, logits.shape[-1]), tgt_out.reshape(-1))
        losses += loss.item()

    return losses / len(list(val_dataloader))
```

现在已经具备了训练模型的所有要素，可以训练模型了。

```
from timeit import default_timer as timer
NUM_EPOCHS = 18

for epoch in range(1, NUM_EPOCHS+1):
    start_time = timer()
    train_loss = train_epoch(transformer, optimizer)
    end_time = timer()
    val_loss = evaluate(transformer)
    print((f"Epoch: {epoch}, Train loss: {train_loss:.3f}, Val loss: {val_loss:.3f}, "f"Epoch time = {(end_time - start_time):.3f}s"))
```

```python
# function to generate output sequence using greedy algorithm
def greedy_decode(model, src, src_mask, max_len, start_symbol):
    src = src.to(DEVICE)
    src_mask = src_mask.to(DEVICE)

    memory = model.encode(src, src_mask)
    ys = torch.ones(1, 1).fill_(start_symbol).type(torch.long).to(DEVICE)
    for i in range(max_len-1):
        memory = memory.to(DEVICE)
        tgt_mask = (generate_square_subsequent_mask(ys.size(0))
                    .type(torch.bool)).to(DEVICE)
        out = model.decode(ys, memory, tgt_mask)
        out = out.transpose(0, 1)
        prob = model.generator(out[:, -1])
        _, next_word = torch.max(prob, dim=1)
        next_word = next_word.item()

        ys = torch.cat([ys,
                        torch.ones(1, 1).type_as(src.data).fill_(next_word)], dim=0)
        if next_word == EOS_IDX:
            break
    return ys


# actual function to translate input sentence into target language
def translate(model: torch.nn.Module, src_sentence: str):
    model.eval()
    src = text_transform[SRC_LANGUAGE](src_sentence).view(-1, 1)
    num_tokens = src.shape[0]
    src_mask = (torch.zeros(num_tokens, num_tokens)).type(torch.bool)
    tgt_tokens = greedy_decode(
        model,  src, src_mask, max_len=num_tokens + 5, start_symbol=BOS_IDX).flatten()
    return " ".join(vocab_transform[TGT_LANGUAGE].lookup_tokens(list(tgt_tokens.cpu().numpy()))).replace("<bos>", "").replace("<eos>", "")
print(translate(transformer, "Eine Gruppe von Menschen steht vor einem Iglu ."))
```

4.5 本章小结

本章介绍了 PyTorch 实现的文本分类，以及构建生成式聊天机器人模型的基本原理，还研究了如何使用 Wav2Vec2ASRBundle 来执行声学特征提取和语音识别。最后介绍了 transformer 实现的机器翻译。

第 5 章 分布式 PyTorch

分布式训练是一种模型训练范式,涉及在多个工作节点上分散训练工作量,从而显著提高训练速度和模型准确性。虽然分布式训练可以用于任何类型的机器学习模型训练,但将其用于大型模型和计算要求高的任务(如深度学习)是最有益的。

5.1 PyTorch 分布式概述

有如下 4 种方法可以在 PyTorch 中执行分布式训练,每种方法在某些用例中都有其优势。
- 分布式数据并行(Distributed Data Parallel,DDP)。
- 全共享数据并行(Fully Sharded Data Parallel,FSDP)。
- 远程过程调用(Remote Procedure Call,RPC)。
- 自定义扩展。

PyTorch 中包含的分布式软件包(即 torc.distributed)使研究人员和从业者能够轻松地跨进程和机器集群进行并行计算。为此,它利用消息传递语义,允许每个进程将数据传递给任何进程。与多处理包(即 torc.multiprocessing)不同,进程可以使用不同的通信后端,并且不限于在同一台机器上执行。

从 PyTorch v1.6.0 开始,torch.distributed 中的功能可以分为如下 3 个主要组件。
- 分布式数据并行训练是一种广泛采用的单程序多数据训练模式。使用 DDP,模型在每个过程中都被复制,每个模型副本都将被提供一组不同的输入数据样本。DDP 负责梯度通信以保持模型副本的同步,并将其与梯度计算重叠以加快训练。
- 基于 RPC 的分布式训练支持无法适应数据并行训练的通用训练结构,如分布式流水线并行性、参数服务器范式,以及 DDP 与其他训练范式的组合。它有助于管理远程对象的生存期,并将 Autograd 引擎扩展到机器边界之外。
- 集体通信库(c10d)支持在组内的进程之间发送张量。它同时提供集体通信 API(如 all_reduce 和 all_gather)和 P2P 通信 API(如 send 和 isend)。DDP 和 RPC 建立在 c10d 上,前者使用集体通信,后者使用 P2P 通信。通常,开发人员不需要直接使用这种原始通信 API,因为 DDP 和 RPC API 可以服务于许多分布式训练场景。然而，在某些用例中，这个 API 仍然是有帮助的。一个例子是分布式参数平均，其中应用程序希望在反向通过之后计算所有模型参数的平均值，而不是使用 DDP 来传达梯度。这可以将通信与计算解耦，并允许对通信内容进行更精细的控制，但另一方面，它也放弃了 DDP 提供的性能优化。

5.1.1 数据并行训练

PyTorch 为数据并行训练提供了多种选择。对于从简单到复杂、从原型到生产的应用程序，常见的开发轨迹如下。

（1）如果数据和模型可以容纳在一个 GPU 中，则使用单设备训练，并且训练速度不是问题。

（2）使用单机多 GPU DataParallel 在一台机器上使用多个 GPU，以最小的代码更改加快训练速度。

（3）如果想进一步加快训练速度并愿意编写更多代码来进行设置，则使用单机多 GPU DistributedDataParallel。

（4）如果应用程序需要跨越机器边界进行扩展，则使用多机器 DistributedDataParallel 和启动脚本。

（5）如果预计会出现错误，可以在训练过程中动态加入和离开，则使用 torch.distributed.elastic 启动分布式训练。

DataParallel 包能够以最低的编码障碍实现单机多 GPU 并行。它只需要对应用程序代码进行一行更改。尽管 DataParallel 非常易于使用，但它通常不会提供最佳性能，因为它在每个前向传递中都复制了模型，而且它的单进程多线程并行性自然会受到 GIL（全局解释器锁）争用的影响。要获得更好的性能，可考虑使用 DistributedDataParallel。

与 DataParallel 相比，DistributedDataParallel 需要多设置一个步骤，即调用 init_process_group。DDP 使用多进程并行，因此模型副本之间不存在 GIL 争用。此外，该模型在 DDP 建造时广播，而不是在每次向前传播时广播，这也有助于加快训练速度。DDP 附带了多种性能优化技术。

随着应用程序复杂性和规模的增长，故障恢复成为一项要求。有时，在使用 DDP 时，不可避免地会遇到内存不足等错误，但 DDP 本身无法从这些错误中恢复，并且不可能使用标准的 try-except 构造来处理这些错误。这是因为 DDP 要求所有进程以紧密同步的方式运行，并且在不同进程中启动的所有 AllReduce 通信必须匹配。如果组中的某个进程引发异常，很可能会导致去同步（AllReduce 操作不匹配），从而导致进程崩溃或挂起。torch.distributed.elastic 增加了容错能力和利用机器动态池的能力。

5.1.2 基于 RPC 的分布式训练

许多训练范式不适合数据并行,如参数服务器范式、分布式流水线并行、具有多个观察者或代理的强化学习应用程序等。torch.distributed.rpc 旨在支持通用的分布式训练场景。

torch.distributed.rpc 有如下 4 个主要支柱。
- RPC 支持在远程工作者上运行给定的函数。
- RRef 有助于管理远程对象的生存周期。
- 分布式 Autograd 将 Autograd 引擎扩展到机器边界之外。
- 分布式优化器会自动联系所有参与的工作者,使用分布式自动梯度引擎计算的梯度来更新参数。

5.2 数据并行

本节将学习如何通过使用 DataParallel 来使用多个 GPU。

在 PyTorch 中使用 GPU 非常容易。可以将模型放在 GPU 上,代码如下:

```
device = torch.device("cuda:0")
model.to(device)
```

然后,可以将所有张量复制到 GPU,代码如下:

```
mytensor = my_tensor.to(device)
```

注意,只需调用 my_tensor.to(device) 即可在 GPU 上返回 my_tensor 的新副本,而不是重写 my_tenstor。需要将其分配给一个新的张量,并在 GPU 上使用该张量。

在多个 GPU 上执行正向、反向传播是很自然的。但是,Pytorch 默认情况下只使用一个 GPU。通过使用 DataParallel 使模型并行运行,可以轻松地在多个 GPU 上运行操作,代码如下:

```
model = nn.DataParallel(model)
```

导入 PyTorch 模块并定义参数,代码如下:

```
import torch
import torch.nn as nn
from torch.utils.data import Dataset, DataLoader

# Parameters and DataLoaders
input_size = 5
output_size = 2
```

```
batch_size = 30
data_size = 100
```

创建设备,代码如下:

```
device = torch.device("cuda:0" if torch.cuda.is_available() else "cpu")
```

制作一个伪数据集,只需要实现 getitem() 方法,代码如下:

```
class RandomDataset(Dataset):

    def __init__(self, size, length):
        self.len = length
        self.data = torch.randn(length, size)

    def __getitem__(self, index):
        return self.data[index]

    def __len__(self):
        return self.len

rand_loader = DataLoader(dataset=RandomDataset(input_size, data_size),
                         batch_size=batch_size, shuffle=True)
```

对于这个演示,模型只是得到一个输入,执行一个线性运算,并给出一个输出。但是,可以在任何模型(CNN、RNN、Capsule Net 等)上使用 DataParallel。

在模型中放置了一个 print 语句来监控输入和输出张量的大小,代码如下:

```
class Model(nn.Module):
    # Our model

    def __init__(self, input_size, output_size):
        super(Model, self).__init__()
        self.fc = nn.Linear(input_size, output_size)

    def forward(self, input):
        output = self.fc(input)
        print("\tIn Model: input size", input.size(),
              "output size", output.size())

        return output
```

首先,需要创建一个模型实例,并检查是否有多个 GPU。如果有多个 GPU,就可以使用 nn.DataParallel 包装模型。然后通过 model.to(device) 将模型放在 GPU 上,代码如下:

```
model = Model(input_size, output_size)
```

```
if torch.cuda.device_count() > 1:
  print("Let's use", torch.cuda.device_count(), "GPUs!")
  # dim = 0 [30, xxx] -> [10, ...], [10, ...], [10, ...] on 3 GPUs
  model = nn.DataParallel(model)

model.to(device)
```

输出：

```
Let's use 4 GPUs!

DataParallel(
  (module): Model(
    (fc): Linear(in_features=5, out_features=2, bias=True)
  )
)
```

现在可以看到输入和输出张量的大小，代码如下：

```
for data in rand_loader:
    input = data.to(device)
    output = model(input)
    print("Outside: input size", input.size(),
          "output_size", output.size())
```

输出结果如下：

```
        In Model: input size torch.Size([8, 5]) output size torch.Size([8, 2])
        In Model: input size torch.Size([8, 5]) output size torch.Size([8, 2])
        In Model: input size torch.Size([8, 5]) output size torch.Size([8, 2])
        In Model: input size torch.Size([6, 5]) output size torch.Size([6, 2])
Outside: input size torch.Size([30, 5]) output_size torch.Size([30, 2])
        In Model: input size torch.Size([8, 5]) output size torch.Size([8, 2])
        In Model: input size torch.Size([8, 5]) output size torch.Size([8, 2])
        In Model: input size torch.Size([8, 5]) output size torch.Size([8, 2])
        In Model: input size torch.Size([6, 5]) output size torch.Size([6, 2])
Outside: input size torch.Size([30, 5]) output_size torch.Size([30, 2])
        In Model: input size torch.Size([8, 5]) output size torch.Size([8, 2])
        In Model: input size torch.Size([8, 5]) output size torch.Size([8, 2])
        In Model: input size torch.Size([8, 5]) output size torch.Size([8, 2])
        In Model: input size torch.Size([6, 5]) output size torch.Size([6, 2])
Outside: input size torch.Size([30, 5]) output_size torch.Size([30, 2])
        In Model: input size torch.Size([3, 5]) output size torch.Size([3, 2])
        In Model: input size torch.Size([3, 5]) output size torch.Size([3, 2])
        In Model: input size torch.Size([3, 5]) output size torch.Size([3, 2])
        In Model: input size torch.Size([1, 5]) output size torch.Size([1, 2])
Outside: input size torch.Size([10, 5]) output_size torch.Size([10, 2])
```

如果有两个GPU，将看到如下结果：

```
# on 2 GPUs
Let's use 2 GPUs!
    In Model: input size torch.Size([15, 5]) output size torch.Size([15, 2])
    In Model: input size torch.Size([15, 5]) output size torch.Size([15, 2])
Outside: input size torch.Size([30, 5]) output_size torch.Size([30, 2])
    In Model: input size torch.Size([15, 5]) output size torch.Size([15, 2])
    In Model: input size torch.Size([15, 5]) output size torch.Size([15, 2])
Outside: input size torch.Size([30, 5]) output_size torch.Size([30, 2])
    In Model: input size torch.Size([15, 5]) output size torch.Size([15, 2])
    In Model: input size torch.Size([15, 5]) output size torch.Size([15, 2])
Outside: input size torch.Size([30, 5]) output_size torch.Size([30, 2])
    In Model: input size torch.Size([5, 5]) output size torch.Size([5, 2])
    In Model: input size torch.Size([5, 5]) output size torch.Size([5, 2])
Outside: input size torch.Size([10, 5]) output_size torch.Size([10, 2])
```

如果有3个GPU，将看到如下结果：

```
Let's use 3 GPUs!
    In Model: input size torch.Size([10, 5]) output size torch.Size([10, 2])
    In Model: input size torch.Size([10, 5]) output size torch.Size([10, 2])
    In Model: input size torch.Size([10, 5]) output size torch.Size([10, 2])
Outside: input size torch.Size([30, 5]) output_size torch.Size([30, 2])
    In Model: input size torch.Size([10, 5]) output size torch.Size([10, 2])
    In Model: input size torch.Size([10, 5]) output size torch.Size([10, 2])
    In Model: input size torch.Size([10, 5]) output size torch.Size([10, 2])
Outside: input size torch.Size([30, 5]) output_size torch.Size([30, 2])
    In Model: input size torch.Size([10, 5]) output size torch.Size([10, 2])
    In Model: input size torch.Size([10, 5]) output size torch.Size([10, 2])
    In Model: input size torch.Size([10, 5]) output size torch.Size([10, 2])
Outside: input size torch.Size([30, 5]) output_size torch.Size([30, 2])
    In Model: input size torch.Size([4, 5]) output size torch.Size([4, 2])
    In Model: input size torch.Size([4, 5]) output size torch.Size([4, 2])
    In Model: input size torch.Size([2, 5]) output size torch.Size([2, 2])
Outside: input size torch.Size([10, 5]) output_size torch.Size([10, 2])
```

如果有8个GPU，将看到如下结果：

```
Let's use 8 GPUs!
    In Model: input size torch.Size([4, 5]) output size torch.Size([4, 2])
    In Model: input size torch.Size([4, 5]) output size torch.Size([4, 2])
    In Model: input size torch.Size([2, 5]) output size torch.Size([2, 2])
    In Model: input size torch.Size([4, 5]) output size torch.Size([4, 2])
    In Model: input size torch.Size([4, 5]) output size torch.Size([4, 2])
    In Model: input size torch.Size([4, 5]) output size torch.Size([4, 2])
```

```
    In Model: input size torch.Size([4, 5]) output size torch.Size([4, 2])
    In Model: input size torch.Size([4, 5]) output size torch.Size([4, 2])
Outside: input size torch.Size([30, 5]) output_size torch.Size([30, 2])
    In Model: input size torch.Size([4, 5]) output size torch.Size([4, 2])
    In Model: input size torch.Size([4, 5]) output size torch.Size([4, 2])
    In Model: input size torch.Size([4, 5]) output size torch.Size([4, 2])
    In Model: input size torch.Size([4, 5]) output size torch.Size([4, 2])
    In Model: input size torch.Size([4, 5]) output size torch.Size([4, 2])
    In Model: input size torch.Size([2, 5]) output size torch.Size([2, 2])
    In Model: input size torch.Size([4, 5]) output size torch.Size([4, 2])
Outside: input size torch.Size([30, 5]) output_size torch.Size([30, 2])
    In Model: input size torch.Size([4, 5]) output size torch.Size([4, 2])
    In Model: input size torch.Size([4, 5]) output size torch.Size([4, 2])
    In Model: input size torch.Size([4, 5]) output size torch.Size([4, 2])
    In Model: input size torch.Size([4, 5]) output size torch.Size([4, 2])
    In Model: input size torch.Size([4, 5]) output size torch.Size([4, 2])
    In Model: input size torch.Size([2, 5]) output size torch.Size([2, 2])
Outside: input size torch.Size([30, 5]) output_size torch.Size([30, 2])
    In Model: input size torch.Size([2, 5]) output size torch.Size([2, 2])
    In Model: input size torch.Size([2, 5]) output size torch.Size([2, 2])
    In Model: input size torch.Size([2, 5]) output size torch.Size([2, 2])
    In Model: input size torch.Size([2, 5]) output size torch.Size([2, 2])
Outside: input size torch.Size([10, 5]) output_size torch.Size([10, 2])
```

DataParallel 会自动分割数据,并将作业指令发送到多个 GPU 上的多个模型。每个模型完成工作后,DataParallel 会收集并合并结果,然后将结果返回。

5.3 单机模型并行最佳实践

模型并行的思想是将模型的不同子网络放置在不同的设备上,并相应地实现正向方法,在跨设备中间输出。由于模型的一部分在任何单个设备上运行,一组设备可以共同为更大的模型服务。

从一个包含两个线性层的玩具模型开始。要在两个 GPU 上运行此模型,只需将每个线性层放在不同的 GPU 上,并相应地移动输入和中间输出以匹配层设备,代码如下:

```
import torch
import torch.nn as nn
import torch.optim as optim
```

```python
class ToyModel(nn.Module):
    def __init__(self):
        super(ToyModel, self).__init__()
        self.net1 = torch.nn.Linear(10, 10).to('cuda:0')
        self.relu = torch.nn.ReLU()
        self.net2 = torch.nn.Linear(10, 5).to('cuda:1')

    def forward(self, x):
        x = self.relu(self.net1(x.to('cuda:0')))
        return self.net2(x.to('cuda:1'))
```

注意，上面的 ToyModel 看起来与在单个 GPU 上实现它的方式非常相似，除了在适当的设备上放置线性层和张量的 4 个 to(device) 调用。与单 GPU 实现相比，这是模型中唯一需要更改的地方。backward() 和 torch.optim 将自动处理梯度，就像模型在一个 GPU 上一样。调用 loss 函数时，只需要确保标签与输出位于同一设备上。

```python
model = ToyModel()
loss_fn = nn.MSELoss()
optimizer = optim.SGD(model.parameters(), lr=0.001)

optimizer.zero_grad()
outputs = model(torch.randn(20, 10))
labels = torch.randn(20, 5).to('cuda:1')
loss_fn(outputs, labels).backward()
optimizer.step()
```

只需更改几行代码，就可以在多个 GPU 上运行现有的单个 GPU 模块。下面的代码显示了如何将 torchvision.models.resnet50() 分解到两个 GPU。其思想是继承现有的 ResNet 模块，并在构建过程中将层拆分到两个 GPU。然后，覆盖 forward() 方法，通过相应地移动中间输出来缝合两个子网络。

```python
from torchvision.models.resnet import ResNet, Bottleneck

num_classes = 1000


class ModelParallelResNet50(ResNet):
    def __init__(self, *args, **kwargs):
        super(ModelParallelResNet50, self).__init__(
            Bottleneck, [3, 4, 6, 3], num_classes=num_classes, *args, **kwargs)

        self.seq1 = nn.Sequential(
```

```python
            self.conv1,
            self.bn1,
            self.relu,
            self.maxpool,
            
            self.layer1,
            self.layer2
        ).to('cuda:0')
        
        self.seq2 = nn.Sequential(
            self.layer3,
            self.layer4,
            self.avgpool,
        ).to('cuda:1')
        
        self.fc.to('cuda:1')
        
    def forward(self, x):
        x = self.seq2(self.seq1(x).to('cuda:1'))
        return self.fc(x.view(x.size(0), -1))
```

上述实现解决了模型太大而不能放入单个 GPU 的情况下的问题。如果模型适合，它将比在单个 GPU 上运行慢。这是因为，在任何时候，两个 GPU 中只有一个在工作，而另一个则无所事事。由于中间输出需要在第二层和第三层之间从 cuda:0 复制到 cuda:1，因此性能进一步恶化。

下面进行一个实验，以获得执行时间的更定量的视图。在这个实验中，通过运行随机输入和标签来训练 ModelParallelResNet50 和现有的 torchvision.models.resnet50()。训练后，模型不会产生任何有用的预测，但可以合理地了解执行时间。

```python
import torchvision.models as models

num_batches = 3
batch_size = 120
image_w = 128
image_h = 128


def train(model):
    model.train(True)
    loss_fn = nn.MSELoss()
    optimizer = optim.SGD(model.parameters(), lr=0.001)
    
    one_hot_indices = torch.LongTensor(batch_size) \
                           .random_(0, num_classes) \
```

```
                    .view(batch_size, 1)

    for _ in range(num_batches):
        # generate random inputs and labels
        inputs = torch.randn(batch_size, 3, image_w, image_h)
        labels = torch.zeros(batch_size, num_classes) \
                    .scatter_(1, one_hot_indices, 1)

        # run forward pass
        optimizer.zero_grad()
        outputs = model(inputs.to('cuda:0'))

        # run backward pass
        labels = labels.to(outputs.device)
        loss_fn(outputs, labels).backward()
        optimizer.step()
```

上面的 train(model) 方法使用 nn.MSELoss() 作为损失函数，optim.SGD() 作为优化器。使用 timeit 运行 train(model) 方法 10 次，并用标准偏差绘制执行时间。

```
import matplotlib.pyplot as plt
plt.switch_backend('Agg')
import numpy as np
import timeit

num_repeat = 10

stmt = "train(model)"

setup = "model = ModelParallelResNet50()"
mp_run_times = timeit.repeat(
    stmt, setup, number=1, repeat=num_repeat, globals=globals())
mp_mean, mp_std = np.mean(mp_run_times), np.std(mp_run_times)

setup = "import torchvision.models as models;" + \
        "model = models.resnet50(num_classes=num_classes).to('cuda:0')"
rn_run_times = timeit.repeat(
    stmt, setup, number=1, repeat=num_repeat, globals=globals())
rn_mean, rn_std = np.mean(rn_run_times), np.std(rn_run_times)


def plot(means, stds, labels, fig_name):
    fig, ax = plt.subplots()
    ax.bar(np.arange(len(means)), means, yerr=stds,
        align='center', alpha=0.5, ecolor='red', capsize=10, width=0.6)
```

```
        ax.set_ylabel('ResNet50 Execution Time (Second)')
        ax.set_xticks(np.arange(len(means)))
        ax.set_xticklabels(labels)
        ax.yaxis.grid(True)
        plt.tight_layout()
        plt.savefig(fig_name)
        plt.close(fig)

plot([mp_mean, rn_mean],
     [mp_std, rn_std],
     ['Model Parallel', 'Single GPU'],
     'mp_vs_rn.png')
```

结果表明,模型并行实现的执行时间比现有的单 GPU 实现长 4.02/3.75-1=7%。因此,可以得出结论,在 GPU 之间来回复制张量大约有 7% 的开销。还有改进的空间,因为知道两个 GPU 中的一个在整个执行过程中都处于闲置状态。一种选择是将每个批次进一步划分为拆分的管道,这样,当一个拆分到达第二个子网络时,下一个拆分可以被馈送到第一个子网络。通过这种方式,两个连续的拆分可以在两个 GPU 上同时运行。

在下面的实验中,将每 120 个图像批次进一步划分为 20 个图像分割。由于 PyTorch 异步启动 CUDA 操作,因此实现不需要派生多个线程来实现并发。

```
class PipelineParallelResNet50(ModelParallelResNet50):
    def __init__(self, split_size=20, *args, **kwargs):
        super(PipelineParallelResNet50, self).__init__(*args, **kwargs)
        self.split_size = split_size

    def forward(self, x):
        splits = iter(x.split(self.split_size, dim=0))
        s_next = next(splits)
        s_prev = self.seq1(s_next).to('cuda:1')
        ret = []

        for s_next in splits:
            # A. ``s_prev`` runs on ``cuda:1``
            s_prev = self.seq2(s_prev)
            ret.append(self.fc(s_prev.view(s_prev.size(0), -1)))

            # B. ``s_next`` runs on ``cuda:0``, which can run concurrently with A
            s_prev = self.seq1(s_next).to('cuda:1')

        s_prev = self.seq2(s_prev)
        ret.append(self.fc(s_prev.view(s_prev.size(0), -1)))
```

```
        return torch.cat(ret)

setup = "model = PipelineParallelResNet50()"
pp_run_times = timeit.repeat(
    stmt, setup, number=1, repeat=num_repeat, globals=globals())
pp_mean, pp_std = np.mean(pp_run_times), np.std(pp_run_times)

plot([mp_mean, rn_mean, pp_mean],
     [mp_std, rn_std, pp_std],
     ['Model Parallel', 'Single GPU', 'Pipelining Model Parallel'],
     'mp_vs_rn_vs_pp.png')
```

实验结果表明,流水线输入到模型并行的 ResNet50,训练过程的速度大约提高了 3.75/2.51-1=49%,距离理想的 100% 加速还有很长的路要走。由于在流水线并行实现中引入了一个新的参数 split_size,因此尚不清楚新参数如何影响整体训练时间。直观地说,使用小的 split_size 会导致许多微小的 CUDA 内核启动,而使用大的 split_size 会导致第一次和最后一次拆分期间相对较长的空闲时间,两者都不是最优的。对于这个特定的实验,可能有一个最佳的 split_size 配置。可以试着通过使用几个不同的 split_size 值进行实验来找到它。

```
means = []
stds = []
split_sizes = [1, 3, 5, 8, 10, 12, 20, 40, 60]

for split_size in split_sizes:
    setup = "model = PipelineParallelResNet50(split_size=%d)" % split_size
    pp_run_times = timeit.repeat(
        stmt, setup, number=1, repeat=num_repeat, globals=globals())
    means.append(np.mean(pp_run_times))
    stds.append(np.std(pp_run_times))

fig, ax = plt.subplots()
ax.plot(split_sizes, means)
ax.errorbar(split_sizes, means, yerr=stds, ecolor='red', fmt='ro')
ax.set_ylabel('ResNet50 Execution Time (Second)')
ax.set_xlabel('Pipeline Split Size')
ax.set_xticks(split_sizes)
ax.yaxis.grid(True)
plt.tight_layout()
plt.savefig("split_size_tradeoff.png")
plt.close(fig)
```

结果表明,将 split_size 设置为 12 可以获得最快的训练速度,这导致 3.75/2.43-1=54% 的加速。

5.4 分布式数据并行入门

Distributed Data Parallel（DDP）在模块级别实现数据并行，可以在多台机器上运行。使用 DDP 的应用程序应该生成多个进程，并为每个进程创建一个 DDP 实例。DDP 使用在 torch.distributed 包中的集体通信来同步梯度和缓冲区。更具体地说，DDP 为 model.parameters() 给出的每个参数注册一个 Autograd 钩子，当在后向过程中计算出相应的梯度时，钩子将触发。然后 DDP 使用该信号来触发进程之间的梯度同步。

建议使用 DDP 的方法是为每个模型复制副本生成一个进程，其中模型复制副本可以跨越多个设备。DDP 进程可以放在同一台机器上或跨机器放置，但 GPU 设备不能跨进程共享。本节从一个基本的 DDP 用例开始，然后演示更高级的用例，包括检查点模型和将 DDP 与模型并行相结合。

在深入讨论之前，澄清一下为什么尽管增加了复杂性，还是会考虑使用 DistributedDataParallel 而不是 DataParallel。

- DataParallel 是单进程、多线程的，并且只能在一台机器上工作，而 DistributedDataParallel 是多进程的，同时适用于单机和多机训练。DataParallel 通常比 DistributedDataParallel 慢，即使在单台机器上也是如此，这是由于线程之间的 GIL 争用、每次迭代的复制模型，以及分散输入和收集输出带来的额外开销。
- DistributedDataParallel 使用模型并行；DataParallel 此时没有。当 DDP 与模型并行相结合时，每个 DDP 进程将使用模型并行，所有进程将共同使用数据并行。

要创建 DDP 模块，必须首先正确设置进程组，代码如下：

```
import os
import sys
import tempfile
import torch
import torch.distributed as dist
import torch.nn as nn
import torch.optim as optim
import torch.multiprocessing as mp

from torch.nn.parallel import DistributedDataParallel as DDP

# On Windows platform, the torch.distributed package only
# supports Gloo backend, FileStore and TcpStore.
# For FileStore, set init_method parameter in init_process_group
# to a local file. Example as follow:
# init_method="file:///f:/libtmp/some_file"
# dist.init_process_group(
#    "gloo",
#    rank=rank,
```

```
#    init_method=init_method,
#    world_size=world_size)
# For TcpStore, same way as on Linux.

def setup(rank, world_size):
    os.environ['MASTER_ADDR'] = 'localhost'
    os.environ['MASTER_PORT'] = '12355'

    # initialize the process group
    dist.init_process_group("gloo", rank=rank, world_size=world_size)

def cleanup():
    dist.destroy_process_group()
```

现在，创建一个玩具模块，用 DDP 封装它，并为它提供一些伪输入数据。由于 DDP 将等级为 0 的进程的模型状态广播到 DDP 构造函数中的所有其他进程，所以，不必担心不同的 DDP 进程从不同的初始模型参数值开始。

```
class ToyModel(nn.Module):
    def __init__(self):
        super(ToyModel, self).__init__()
        self.net1 = nn.Linear(10, 10)
        self.relu = nn.ReLU()
        self.net2 = nn.Linear(10, 5)

    def forward(self, x):
        return self.net2(self.relu(self.net1(x)))


def demo_basic(rank, world_size):
    print(f"Running basic DDP example on rank {rank}.")
    setup(rank, world_size)

    # create model and move it to GPU with id rank
    model = ToyModel().to(rank)
    ddp_model = DDP(model, device_ids=[rank])

    loss_fn = nn.MSELoss()
    optimizer = optim.SGD(ddp_model.parameters(), lr=0.001)

    optimizer.zero_grad()
    outputs = ddp_model(torch.randn(20, 10))
    labels = torch.randn(20, 5).to(rank)
    loss_fn(outputs, labels).backward()
    optimizer.step()
```

```
    cleanup()


def run_demo(demo_fn, world_size):
    mp.spawn(demo_fn,
             args=(world_size,),
             nprocs=world_size,
             join=True)
```

DDP 包装了较低级别的分布式通信细节,并提供了一个干净的 API,就好像它是一个本地模型一样。梯度同步通信发生在反向通过期间,与反向计算重叠。当 backward() 返回时,param.grad 已经包含同步梯度张量。

在 DDP 中,构造函数、前向传递和后向传递是分布式同步点。不同的进程预计将启动相同数量的同步,并以相同的顺序到达这些同步点,并在大致相同的时间进入每个同步点。否则,快速进程可能会提前到达并在等待掉队者时超时。因此,用户负责平衡跨进程的工作负载分布。有时,由于网络延迟、资源争用或不可预测的工作负载峰值等原因,处理速度出现偏差是不可避免的。为了避免在这些情况下超时,应确保在调用 init_process_group 时传递足够大的超时值。

在训练和从检查点恢复过程中,使用 torch.save 和 torch.load 到检查点模块是很常见的。当使用 DDP 时,一种优化是只在一个进程中保存模型,然后将其加载到所有进程,从而减少写入开销。这是正确的,因为所有过程都从相同的参数开始,并且梯度在反向过程中是同步的,所以,优化器应该将参数设置为相同的值。如果使用此优化,应确保在保存完成之前没有进程开始加载。此外,在加载模块时,需要提供适当的 map_location 参数,以防止进程进入其他进程的设备。如果缺少 map_location,torch.load 将首先将模块加载到 CPU,然后将每个参数复制到保存的位置,这将导致同一台机器上的所有进程使用相同的设备集。

```
def demo_checkpoint(rank, world_size):
    print(f"Running DDP checkpoint example on rank {rank}.")
    setup(rank, world_size)

    model = ToyModel().to(rank)
    ddp_model = DDP(model, device_ids=[rank])

    CHECKPOINT_PATH = tempfile.gettempdir() + "/model.checkpoint"
    if rank == 0:
        # All processes should see same parameters as they all start from same
        # random parameters and gradients are synchronized in backward passes.
        # Therefore, saving it in one process is sufficient.
```

```
        torch.save(ddp_model.state_dict(), CHECKPOINT_PATH)

    # Use a barrier() to make sure that process 1 loads the model after process
    # 0 saves it.
    dist.barrier()
    # configure map_location properly
    map_location = {'cuda:%d' % 0: 'cuda:%d' % rank}
    ddp_model.load_state_dict(
        torch.load(CHECKPOINT_PATH, map_location=map_location))

    loss_fn = nn.MSELoss()
    optimizer = optim.SGD(ddp_model.parameters(), lr=0.001)

    optimizer.zero_grad()
    outputs = ddp_model(torch.randn(20, 10))
    labels = torch.randn(20, 5).to(rank)

    loss_fn(outputs, labels).backward()
    optimizer.step()

    # Not necessary to use a dist.barrier() to guard the file deletion below
    # as the AllReduce ops in the backward pass of DDP already served as
    # a synchronization.

    if rank == 0:
        os.remove(CHECKPOINT_PATH)

    cleanup()
```

DDP 也适用于多 GPU 模型。DDP 封装多 GPU 模型在训练具有大量数据的大型模型时尤其有用。

```
class ToyMpModel(nn.Module):
    def __init__(self, dev0, dev1):
        super(ToyMpModel, self).__init__()
        self.dev0 = dev0
        self.dev1 = dev1
        self.net1 = torch.nn.Linear(10, 10).to(dev0)
        self.relu = torch.nn.ReLU()
        self.net2 = torch.nn.Linear(10, 5).to(dev1)

    def forward(self, x):
        x = x.to(self.dev0)
        x = self.relu(self.net1(x))
        x = x.to(self.dev1)
```

```
        return self.net2(x)
```

将多 GPU 模型传递给 DDP 时，必须不设置 device_ids 和 output_device。输入和输出数据将通过应用程序或模型 forward() 方法放置在适当的设备中。

```
def demo_model_parallel(rank, world_size):
    print(f"Running DDP with model parallel example on rank {rank}.")
    setup(rank, world_size)

    # setup mp_model and devices for this process
    dev0 = rank * 2
    dev1 = rank * 2 + 1
    mp_model = ToyMpModel(dev0, dev1)
    ddp_mp_model = DDP(mp_model)

    loss_fn = nn.MSELoss()
    optimizer = optim.SGD(ddp_mp_model.parameters(), lr=0.001)

    optimizer.zero_grad()
    # outputs will be on dev1
    outputs = ddp_mp_model(torch.randn(20, 10))
    labels = torch.randn(20, 5).to(dev1)
    loss_fn(outputs, labels).backward()
    optimizer.step()

    cleanup()


if __name__ == "__main__":
    n_gpus = torch.cuda.device_count()
    assert n_gpus >= 2, f"Requires at least 2 GPUs to run, but got {n_gpus}"
    world_size = n_gpus
    run_demo(demo_basic, world_size)
    run_demo(demo_checkpoint, world_size)
    world_size = n_gpus//2
    run_demo(demo_model_parallel, world_size)
```

可以利用 PyTorch Elastic 来简化 DDP 代码，并更容易地初始化作业。下面仍然使用 Toymodel 示例，并创建一个名为 elastic_ddp.py 的文件。

```
import torch
import torch.distributed as dist
import torch.nn as nn
import torch.optim as optim

from torch.nn.parallel import DistributedDataParallel as DDP
```

```python
class ToyModel(nn.Module):
    def __init__(self):
        super(ToyModel, self).__init__()
        self.net1 = nn.Linear(10, 10)
        self.relu = nn.ReLU()
        self.net2 = nn.Linear(10, 5)

    def forward(self, x):
        return self.net2(self.relu(self.net1(x)))


def demo_basic():
    dist.init_process_group("nccl")
    rank = dist.get_rank()
    print(f"Start running basic DDP example on rank {rank}.")

    # create model and move it to GPU with id rank
    device_id = rank % torch.cuda.device_count()
    model = ToyModel().to(device_id)
    ddp_model = DDP(model, device_ids=[device_id])

    loss_fn = nn.MSELoss()
    optimizer = optim.SGD(ddp_model.parameters(), lr=0.001)

    optimizer.zero_grad()
    outputs = ddp_model(torch.randn(20, 10))
    labels = torch.randn(20, 5).to(device_id)
    loss_fn(outputs, labels).backward()
    optimizer.step()
    dist.destroy_process_group()

if __name__ == "__main__":
    demo_basic()
```

然后可以在所有节点上运行 torchrun 命令，以初始化上面创建的 DDP 作业。

```
torchrun --nnodes=2 --nproc_per_node=8 --rdzv_id=100 --rdzv_backend=c10d --rdzv_endpoint=$MASTER_ADDR:29400 elastic_ddp.py
```

在两个主机上运行 DDP 脚本，每个主机运行 8 个进程，也就是说，在 16 个 GPU 上运行它。注意，$MASTER_ADDR 在所有节点上必须相同。

在这里，torchrun 将启动 8 个进程，并在其启动的节点上的每个进程上调用 elastic_ddp.py，但用户还需要应用像 SLURM 这样的集群管理工具才能在两个节点上实际运行此命令。

例如，在启用 SLURM 的集群上，可以编写一个脚本来运行上面的命令，并对 MASTER_ADDR 进行如下设置：

```
export MASTER_ADDR=$(scontrol show hostname ${SLURM_NODELIST} | head -n 1)
```

然后可以使用 SLURM 命令运行这个脚本：srun --nodes=2 ./torchrun_script.sh。当然，这只是一个例子，可以选择自己的集群调度工具来启动 torchrun 作业。

5.5 用 PyTorch 编写分布式应用程序

编写分布式程序，需要能够同时运行多个进程。如果有权访问计算集群，则应该向本地系统管理员查询，或者使用喜欢的协调工具（例如，pdsh、clustershell 或其他工具）。本节将使用一台机器，并使用以下模板生成多个进程。

```python
"""run.py:"""
#!/usr/bin/env python
import os
import torch
import torch.distributed as dist
import torch.multiprocessing as mp

def run(rank, size):
    """ Distributed function to be implemented later. """
    pass

def init_process(rank, size, fn, backend='gloo'):
    """ Initialize the distributed environment. """
    os.environ['MASTER_ADDR'] = '127.0.0.1'
    os.environ['MASTER_PORT'] = '29500'
    dist.init_process_group(backend, rank=rank, world_size=size)
    fn(rank, size)


if __name__ == "__main__":
    size = 2
    processes = []
    mp.set_start_method("spawn")
    for rank in range(size):
        p = mp.Process(target=init_process, args=(rank, size, run))
        p.start()
        processes.append(p)
```

```
    for p in processes:
        p.join()
```

上面的脚本产生了两个进程,每个进程将设置分布式环境,初始化进程组(dist.init_process_group),并最终执行给定的 run 函数。

init_process() 函数确保每个进程都能通过使用相同的 IP 地址和端口的主进程进行协调。注意,这里使用了 Gloo 后端,其他后端也可用。

从一个进程到另一个进程的数据传输称为点对点通信。这些是通过 send 和 recv 函数实现的。

```
"""Blocking point-to-point communication."""

def run(rank, size):
    tensor = torch.zeros(1)
    if rank == 0:
        tensor += 1
        # Send the tensor to process 1
        dist.send(tensor=tensor, dst=1)
    else:
        # Receive tensor from process 0
        dist.recv(tensor=tensor, src=0)
    print('Rank ', rank, ' has data ', tensor[0])
```

在上面的例子中,两个进程都以零张量开始,然后进程 0 递增张量并将其发送到进程 1,使它们都以 2.0 结束。注意,进程 1 需要分配内存,以便存储将要接收的数据。

还要注意 send/recv 正在阻塞:两个进程都会停止,直到通信完成。另一方面,即时消息是非阻塞的;脚本继续执行,方法返回一个 Work 对象,可以选择 wait()。

```
"""Non-blocking point-to-point communication."""

def run(rank, size):
    tensor = torch.zeros(1)
    req = None
    if rank == 0:
        tensor += 1
        # Send the tensor to process 1
        req = dist.isend(tensor=tensor, dst=1)
        print('Rank 0 started sending')
    else:
        # Receive tensor from process 0
        req = dist.irecv(tensor=tensor, src=0)
        print('Rank 1 started receiving')
    req.wait()
    print('Rank ', rank, ' has data ', tensor[0])
```

当使用即时性函数时，必须小心如何使用发送和接收张量。由于不知道数据何时会被传递到另一个进程，所以，在 req.wait() 完成之前，不能修改发送的张量，也不能访问接收的张量。换句话说，
- 在 dist.isend() 之后写入张量将导致未定义的行为。
- 在 dist.irecv() 之后读取张量将导致未定义的行为。

在执行 req.wait() 之后，可以保证通信已经发生，并且存储在 tensor[0] 中的值是 1.0。
当希望对进程的通信进行更细粒度的控制时，点对点通信非常有用。

与点对点通信相反，集体通信允许在一个组中的所有进程之间建立通信模式。组是所有进程的子集。要创建一个组，可以将一个等级列表传递给 dist.new_group(group)。默认情况下，集体通信在所有进程上执行。例如，为了获得所有进程上所有张量的和，可以使用 dist.all_reduce(tensor, op, group) 进行集体通信。

```
""" All-Reduce example."""
def run(rank, size):
    """ Simple collective communication. """
    group = dist.new_group([0, 1])
    tensor = torch.ones(1)
    dist.all_reduce(tensor, op=dist.ReduceOp.SUM, group=group)
    print('Rank ', rank, ' has data ', tensor[0])
```

想要组中所有张量的和，可以使用 dist.ReduceOp.SUM 作为 reduce 运算符。一般来说，任何可交换的数学运算都可以用作运算符。PyTorch 配备了 4 个开箱即用的运算符，它们都在元素级别工作。
- dist.ReduceOp.SUM。
- dist.ReduceOp.PRODUCT。
- dist.ReduceOp.MAX。
- dist.ReduceOp.MIN。

除了 dist.all_reduce(tensor, op, group)，PyTorch 中目前共有 7 个集体通信函数。
- dist.broadcast(tensor, src, group)：将 tensor 从 src 复制到所有其他进程。
- dist.reduce(tensor, dst, op, group)：将 op 应用于每个 tensor 并将结果存储在 dst 中。
- dist.all_reduce(tensor, op, group)：与 reduce 相同，但结果存储在所有进程中。
- dist.scatter(tensor, scatter_list, src, group)：复制第 i 个张量 scatter_list[i] 到第 i 个进程。
- dist.gather(tensor, gather_list, dst, group)：复制 dst 中所有进程的 tensor。
- dist.all_gather(tensor_list, tensor, group)：将所有进程的 tensor 复制到所有进程上的 tensor_list。
- dist.barrier(group)：阻止组中的所有进程，直到每个进程都进入该函数。

既然了解了分布式模块是如何工作的，那么就用它编写一些有用的东西，目标是复制

DistributedDataParallel 的功能。

很简单，我们想要实现随机梯度下降的分布式版本。脚本将允许所有进程在其一批数据上计算其模型的梯度，然后对其梯度进行平均。为了在改变进程数量时确保类似的收敛结果，首先必须对数据集进行分区。

```python
""" Dataset partitioning helper """
class Partition(object):

    def __init__(self, data, index):
        self.data = data
        self.index = index

    def __len__(self):
        return len(self.index)

    def __getitem__(self, index):
        data_idx = self.index[index]
        return self.data[data_idx]


class DataPartitioner(object):

    def __init__(self, data, sizes=[0.7, 0.2, 0.1], seed=1234):
        self.data = data
        self.partitions = []
        rng = Random()
        rng.seed(seed)
        data_len = len(data)
        indexes = [x for x in range(0, data_len)]
        rng.shuffle(indexes)

        for frac in sizes:
            part_len = int(frac * data_len)
            self.partitions.append(indexes[0:part_len])
            indexes = indexes[part_len:]

    def use(self, partition):
        return Partition(self.data, self.partitions[partition])
```

有了上面的代码片段，现在可以使用以下几行代码简单地对任何数据集进行分区。

```python
""" Partitioning MNIST """
def partition_dataset():
    dataset = datasets.MNIST('./data', train=True, download=True,
                             transform=transforms.Compose([
```

```
                            transforms.ToTensor(),
                            transforms.Normalize((0.1307,), (0.3081,))
                        ]))
    size = dist.get_world_size()
    bsz = 128 / float(size)
    partition_sizes = [1.0 / size for _ in range(size)]
    partition = DataPartitioner(dataset, partition_sizes)
    partition = partition.use(dist.get_rank())
    train_set = torch.utils.data.DataLoader(partition,
                                            batch_size=bsz,
                                            shuffle=True)
    return train_set, bsz
```

假设有两个副本，那么每个进程将有一个 60000/2=30000 个样本的训练集。并将批量大小除以副本数量，以保持 128 的总批量大小。

现在可以编写通常的前向 - 后向优化训练代码，并添加一个函数调用来平均模型的梯度。

```
""" Distributed Synchronous SGD Example """
def run(rank, size):
    torch.manual_seed(1234)
    train_set, bsz = partition_dataset()
    model = Net()
    optimizer = optim.SGD(model.parameters(),
                          lr=0.01, momentum=0.5)

    num_batches = ceil(len(train_set.dataset) / float(bsz))
    for epoch in range(10):
        epoch_loss = 0.0
        for data, target in train_set:
            optimizer.zero_grad()
            output = model(data)
            loss = F.nll_loss(output, target)
            epoch_loss += loss.item()
            loss.backward()
            average_gradients(model)
            optimizer.step()
        print('Rank ', dist.get_rank(), ', epoch ',
              epoch, ': ', epoch_loss / num_batches)
```

仍然需要实现 average_gradients(model) 函数，该函数只需接收一个模型并对其在整个世界中的梯度进行平均。

```
""" Gradient averaging. """
def average_gradients(model):
```

```
        size = float(dist.get_world_size())
        for param in model.parameters():
            dist.all_reduce(param.grad.data, op=dist.ReduceOp.SUM)
            param.grad.data /= size
```

成功地实现了分布式同步 SGD，并且可以在大型计算机集群上训练任何模型。

现在已准备好探索 torch.distributed 的一些更高级的功能。下面主要介绍下面两种功能：

（1）通信后端：学习如何使用 MPI 和 Gloo 进行 GPU-GPU 通信。

（2）初始化方法：了解如何在 dist.init_process_group() 中优雅地设置初始协调阶段。

torch.distributed 最优雅的几点之一是能够在不同的后端上进行抽象和构建。PyTorch 中目前实现了 3 个后端：Gloo、NCCL 和 MPI。它们都有不同的规格和权衡，这取决于所需的用例。

到目前为止，已经广泛使用了 Gloo 后端。作为一个开发平台，它非常方便，因为它包含在预编译的 PyTorch 二进制文件中，并且可以在 Linux（自 0.2 版起）和 macOS（自 1.3 版起）上工作。它支持 CPU 上的所有点对点和集体操作，以及 GPU 上的所有集体操作。CUDA 张量的集体操作的实现不如 NCCL 后端提供的那样优化。

如果将模型放在 GPU 上，分布式 SGD 示例将不起作用。为了使用多个 GPU，还需要进行以下修改：

（1）使用 device = torch.device("cuda:{}".format(rank))。

（2）model = Net() → model = Net().to(device)。

（3）使用 data, target = data.to(device), target.to(device)。

经过上述修改，模型现在正在两个 GPU 上进行训练，可以使用 watch nvidia-smi 监控它们的利用率。

消息传递接口（Message Passing Interface，MPI）是高性能计算领域的标准化工具。它允许进行点对点和集体通信，是 torch.distributed 的 API 的主要灵感来源。MPI 有多种实现方式（如 Open MPI、MVAPICH2、Intel MPI），每种都针对不同的目的进行了优化。使用 MPI 后端的优势在于 MPI 在大型计算机集群上具有广泛的可用性和高级别的优化。

不幸的是，PyTorch 的二进制文件不能包含 MPI 实现，因此，不得不手动重新编译它。幸运的是，这个过程相当简单，因为在编译时，PyTorch 将自己寻找可用的 MPI 实现。

为了测试新安装的后端，需要进行一些修改。

（1）将 if __name__ == '__main__': 下的内容替换为 init_process(0, 0, run, backend='mpi')。

（2）运行 mpirun -n 4 python myscript.py。

这些更改的原因是 MPI 需要在生成进程之前创建自己的环境。MPI 还将生成自己的进程，并执行初始化方法中描述的握手，从而使 init_process_group 的 rank 和 size 参数变得多余。这实际上非常强大，因为可以向 mpirun 传递额外的参数，以便为每个进程定制计算资源。这样做以后，应该获得与其他通信后端相同的熟悉输出。

NCCL 后端提供了针对 CUDA 张量的集体操作的优化实现。如果只在集体操作中使用 CUDA 张量，则可以考虑使用此后端以获得最佳性能。NCCL 后端包含在具有 CUDA 支持的预构建二进制文件中。

下面介绍调用的第一个函数：dist.init_process_group(backend, init_method)。将介绍不同的初始化方法，这些方法负责每个进程之间的初始协调步骤。这些方法允许定义如何进行协调。

本节一直在使用环境变量初始化方法。通过在所有机器上设置以下 4 个环境变量，所有进程都将能够正确连接到主进程，获得有关其他进程的信息，并最终与它们握手。

- MASTER_PORT：计算机上的一个空闲端口，用于承载等级为 0 的进程。
- MASTER_ADDR：将托管等级为 0 的进程的计算机的 IP 地址。
- WORLD_SIZE：进程的总数，以便主进程知道要等待多少工作进程。
- RANK：每个进程的等级，这样它们就会知道它是工作者还是主人。

共享文件系统要求所有进程都可以访问共享文件系统，并通过共享文件对它们进行协调。这意味着每个进程都将打开文件，写入其信息，并等待所有进程都这样做。之后，所有进程都可以随时获得所需的所有信息。为了避免竞态条件，文件系统必须支持通过 fcntl 进行锁定。

```
dist.init_process_group(
    init_method='file:///mnt/nfs/sharedfile',
    rank=args.rank,
    world_size=4)
```

通常 TCP 初始化的方法可以通过提供级别为 0 的进程的 IP 地址和可访问的端口号来实现。在这里，所有工作者都可以连接到级别为 0 的进程，并交换如何联系彼此的信息。

```
dist.init_process_group(
    init_method='tcp://10.1.1.20:23456',
    rank=args.rank,
    world_size=4)
```

5.6 完全分片数据并行入门

大规模训练人工智能模型是一项具有挑战性的任务，需要大量的计算能力和资源。处理这些非常大的模型的训练也具有相当大的工程复杂性。在 PyTorch 1.11 中发布的 PyTorch FSDP（Fully Sharded Data Parallel），使大规模训练更容易。

本节将介绍如何使用 FSDP API，用于简单的 MNIST 模型，这些模型可以扩展到其他更大的模型，如 HuggingFace BERT 模型、GPT 3 模型，最高可达 1T 参数。

在 Distributed Data Parallel（DDP）训练中，每个进程／工作者都拥有一个模型的副本并处理一批数据，最后使用 all-reduce 来汇总不同工作者的梯度。在 DDP 中，模型权重和优化器状态在所有工作者中复制。FSDP 是一种数据并行性，可在 DDP 等级之间对模型参数、优化器状态和梯度进行分片。

FSDP GPU 在所有工作者中的内存占用将小于 DDP，这使得一些非常大的模型的训练变得可行，并有助于为训练工作适应更大的模型或批量，这将伴随着通信量增加的成本。可以通过诸如通信和计算重叠之类的内部优化来减少通信开销。

FSDP 的工作原理如下：

在构造函数中：

- 分片模型参数和每个等级只保留自己的分片。

在正向路径中：

- 运行 all_gather 从所有等级中收集所有分片，以恢复该 FSDP 单元中的完整参数。
- 运行前向计算。
- 丢弃刚刚收集的参数分片。

在后向路径中：

- 运行 all_gather 从所有等级中收集所有分片，以恢复该 FSDP 单元中的完整参数。
- 运行后向计算。
- 运行 reduce_scatter 以同步梯度。
- 丢弃参数。

在这里，使用一个玩具模型在 MNIST 数据集上运行训练以进行演示。类似地，API 和逻辑可以应用于更大的模型进行训练。

安装 PyTorch 和 Torchvision，代码如下：

```
pip3 install --pre torch torchvision torchaudio -f https://download.pytorch.org/whl/nightly/cu113/torch_nightly.html
```

将以下代码片段添加到 Python 脚本 "FSDP_mnist.py" 中。

导入必要的包，代码如下：

```
# Based on: https://github.com/pytorch/examples/blob/master/mnist/main.py
import os
import argparse
import functools
import torch
import torch.nn as nn
import torch.nn.functional as F
import torch.optim as optim
```

```python
from torchvision import datasets, transforms

from torch.optim.lr_scheduler import StepLR

import torch.distributed as dist
import torch.multiprocessing as mp
from torch.nn.parallel import DistributedDataParallel as DDP
from torch.utils.data.distributed import DistributedSampler
from torch.distributed.fsdp import FullyShardedDataParallel as FSDP
from torch.distributed.fsdp.fully_sharded_data_parallel import (
    CPUOffload,
    BackwardPrefetch,
)
from torch.distributed.fsdp.wrap import (
    size_based_auto_wrap_policy,
    enable_wrap,
    wrap,
)
```

下面进行分布式训练设置。FSDP 是一种数据并行，需要分布式训练环境，因此，这里使用两个辅助函数来初始化分布式训练和清理的过程，代码如下：

```python
def setup(rank, world_size):
    os.environ['MASTER_ADDR'] = 'localhost'
    os.environ['MASTER_PORT'] = '12355'

    # initialize the process group
    dist.init_process_group("nccl", rank=rank, world_size=world_size)

def cleanup():
    dist.destroy_process_group()
```

为手写数字分类定义玩具模型，代码如下：

```python
class Net(nn.Module):
    def __init__(self):
        super(Net, self).__init__()
        self.conv1 = nn.Conv2d(1, 32, 3, 1)
        self.conv2 = nn.Conv2d(32, 64, 3, 1)
        self.dropout1 = nn.Dropout(0.25)
        self.dropout2 = nn.Dropout(0.5)
        self.fc1 = nn.Linear(9216, 128)
        self.fc2 = nn.Linear(128, 10)

    def forward(self, x):
```

```python
        x = self.conv1(x)
        x = F.relu(x)
        x = self.conv2(x)
        x = F.relu(x)
        x = F.max_pool2d(x, 2)
        x = self.dropout1(x)
        x = torch.flatten(x, 1)
        x = self.fc1(x)
        x = F.relu(x)
        x = self.dropout2(x)
        x = self.fc2(x)
        output = F.log_softmax(x, dim=1)
        return output
```

定义训练函数,代码如下:

```python
def train(args, model, rank, world_size, train_loader, optimizer, epoch, sampler=None):
    model.train()
    ddp_loss = torch.zeros(2).to(rank)
    if sampler:
        sampler.set_epoch(epoch)
    for batch_idx, (data, target) in enumerate(train_loader):
        data, target = data.to(rank), target.to(rank)
        optimizer.zero_grad()
        output = model(data)
        loss = F.nll_loss(output, target, reduction='sum')
        loss.backward()
        optimizer.step()
        ddp_loss[0] += loss.item()
        ddp_loss[1] += len(data)

    dist.all_reduce(ddp_loss, op=dist.ReduceOp.SUM)
    if rank == 0:
        print('Train Epoch: {} \tLoss: {:.6f}'.format(epoch, ddp_loss[0] / ddp_loss[1]))
```

定义验证函数,代码如下:

```python
def test(model, rank, world_size, test_loader):
    model.eval()
    correct = 0
    ddp_loss = torch.zeros(3).to(rank)
    with torch.no_grad():
        for data, target in test_loader:
            data, target = data.to(rank), target.to(rank)
```

```python
            output = model(data)
            ddp_loss[0] += F.nll_loss(output, target, reduction='sum').item()  # sum up batch loss
            pred = output.argmax(dim=1, keepdim=True)  # get the index of the max log-probability
            ddp_loss[1] += pred.eq(target.view_as(pred)).sum().item()
            ddp_loss[2] += len(data)

    dist.all_reduce(ddp_loss, op=dist.ReduceOp.SUM)

    if rank == 0:
        test_loss = ddp_loss[0] / ddp_loss[2]
        print('Test set: Average loss: {:.4f}, Accuracy: {}/{} ({:.2f}%)\n'.format(
            test_loss, int(ddp_loss[1]), int(ddp_loss[2]),
            100. * ddp_loss[1] / ddp_loss[2]))
```

定义将模型封装在 FSDP 中的分布式训练函数，代码如下：

```python
def fsdp_main(rank, world_size, args):
    setup(rank, world_size)

    transform=transforms.Compose([
        transforms.ToTensor(),
        transforms.Normalize((0.1307,), (0.3081,))
    ])

    dataset1 = datasets.MNIST('../data', train=True, download=True,
                        transform=transform)
    dataset2 = datasets.MNIST('../data', train=False,
                        transform=transform)

    sampler1 = DistributedSampler(dataset1, rank=rank, num_replicas=world_size, shuffle=True)
    sampler2 = DistributedSampler(dataset2, rank=rank, num_replicas=world_size)

    train_kwargs = {'batch_size': args.batch_size, 'sampler': sampler1}
    test_kwargs = {'batch_size': args.test_batch_size, 'sampler': sampler2}
    cuda_kwargs = {'num_workers': 2,
                    'pin_memory': True,
                    'shuffle': False}
    train_kwargs.update(cuda_kwargs)
    test_kwargs.update(cuda_kwargs)

    train_loader = torch.utils.data.DataLoader(dataset1,**train_kwargs)
    test_loader = torch.utils.data.DataLoader(dataset2, **test_kwargs)
```

```python
        my_auto_wrap_policy = functools.partial(
            size_based_auto_wrap_policy, min_num_params=100
        )
        torch.cuda.set_device(rank)


        init_start_event = torch.cuda.Event(enable_timing=True)
        init_end_event = torch.cuda.Event(enable_timing=True)

        model = Net().to(rank)

        model = FSDP(model)

        optimizer = optim.Adadelta(model.parameters(), lr=args.lr)

        scheduler = StepLR(optimizer, step_size=1, gamma=args.gamma)
        init_start_event.record()
        for epoch in range(1, args.epochs + 1):
            train(args, model, rank, world_size, train_loader, optimizer, epoch, sampler=sampler1)
            test(model, rank, world_size, test_loader)
            scheduler.step()

        init_end_event.record()

        if rank == 0:
            print(f"CUDA event elapsed time: {init_start_event.elapsed_time(init_end_event) / 1000}sec")
            print(f"{model}")

        if args.save_model:
            # use a barrier to make sure training is done on all ranks
            dist.barrier()
            # state_dict for FSDP model is only available on Nightlies for now
            states = model.state_dict()
            if rank == 0:
                torch.save(states, "mnist_cnn.pt")

        cleanup()
```

最后解析参数并设置主函数，代码如下：

```python
if __name__ == '__main__':
    # Training settings
    parser = argparse.ArgumentParser(description='PyTorch MNIST Example')
```

```python
parser.add_argument('--batch-size', type=int, default=64, metavar='N',
                    help='input batch size for training (default: 64)')
parser.add_argument('--test-batch-size', type=int, default=1000, metavar='N',
                    help='input batch size for testing (default: 1000)')
parser.add_argument('--epochs', type=int, default=10, metavar='N',
                    help='number of epochs to train (default: 14)')
parser.add_argument('--lr', type=float, default=1.0, metavar='LR',
                    help='learning rate (default: 1.0)')
parser.add_argument('--gamma', type=float, default=0.7, metavar='M',
                    help='Learning rate step gamma (default: 0.7)')
parser.add_argument('--no-cuda', action='store_true', default=False,
                    help='disables CUDA training')
parser.add_argument('--seed', type=int, default=1, metavar='S',
                    help='random seed (default: 1)')
parser.add_argument('--save-model', action='store_true', default=False,
                    help='For Saving the current Model')
args = parser.parse_args()

torch.manual_seed(args.seed)

WORLD_SIZE = torch.cuda.device_count()
mp.spawn(fsdp_main,
    args=(WORLD_SIZE, args),
    nprocs=WORLD_SIZE,
    join=True)
```

用 FSDP 包裹模型，模型将如下所示，可以看到模型已经包裹在一个 FSDP 单元中。接下来将考虑添加 fsdp_auto_wrap_policy，并讨论其中的差异。

```
FullyShardedDataParallel(
(_fsdp_wrapped_module): FlattenParamsWrapper(
    (_fpw_module): Net(
    (conv1): Conv2d(1, 32, kernel_size=(3, 3), stride=(1, 1))
    (conv2): Conv2d(32, 64, kernel_size=(3, 3), stride=(1, 1))
    (dropout1): Dropout(p=0.25, inplace=False)
    (dropout2): Dropout(p=0.5, inplace=False)
    (fc1): Linear(in_features=9216, out_features=128, bias=True)
    (fc2): Linear(in_features=128, out_features=10, bias=True)
    )
)
)
```

在 FSDP 中应用 fsdp_auto_wrap_policy，否则，FSDP 将把整个模型放在一个 FSDP 单元中，这将降低计算效率和内存效率。其工作原理如下：假设模型包含 100 个线性层，如果执行 FSDP（模型），将只有一个 FSDP 单元来包装整个模型。在这种情况下，allgather

将收集所有 100 个线性层的完整参数，因此，不会为参数分片节省 CUDA 内存。此外，对于所有 100 个线性层，只有一个阻塞 allgather 调用，层之间不会有通信和计算重叠。

为了避免这种情况，可以传入一个 fsdp_auto_wrap_policy，它将密封当前 FSDP 单元，并在满足指定条件（如大小限制）时自动启动一个新单元。通过这种方式，将拥有多个 FSDP 单元，并且一次只需要一个 FSDP 单元来收集完整的参数。例如，假设有 5 个 FSDP 单元，每个单元包裹 20 个线性层。在前向中，第一个 FSDP 单元将收集前 20 个线性层的所有参数，进行计算，丢弃这些参数，然后继续到接下来的 20 个线性层。因此，在任何时间点，每个级别只具体化 20 个线性层的参数 / 梯度，而不是 100 个。

为了做到这一点，定义了 auto_wrap_policy 并将其传递给 FSDP 包装器，在下面的示例中，my_auto_wrap_policy 定义了如果一个层中的参数数量大于 100，则该层可以由 FSDP 包装或分片。如果该层中的参数数量小于 100，则会通过 FSDP 将其与其他小层包裹在一起。找到一个最佳的自动包装策略是一项挑战，PyTorch 将在未来为此配置添加自动调优。在没有自动调优工具的情况下，最好通过实验使用不同的自动包装策略来评测工作流，并找到最佳策略。

```
my_auto_wrap_policy = functools.partial(
    size_based_auto_wrap_policy, min_num_params=20000
)
torch.cuda.set_device(rank)
model = Net().to(rank)

model = FSDP(model,
    fsdp_auto_wrap_policy=my_auto_wrap_policy)
```

应用 FSDP_auto_wrap_policy 后的模型如下：

```
  FullyShardedDataParallel(
(_fsdp_wrapped_module): FlattenParamsWrapper(
  (_fpw_module): Net(
    (conv1): Conv2d(1, 32, kernel_size=(3, 3), stride=(1, 1))
    (conv2): Conv2d(32, 64, kernel_size=(3, 3), stride=(1, 1))
    (dropout1): Dropout(p=0.25, inplace=False)
    (dropout2): Dropout(p=0.5, inplace=False)
    (fc1): FullyShardedDataParallel(
      (_fsdp_wrapped_module): FlattenParamsWrapper(
        (_fpw_module): Linear(in_features=9216, out_features=128, bias=True)
      )
    )
    (fc2): Linear(in_features=128, out_features=10, bias=True)
  )
)
```

5.7 基于完全分片数据并行的高级模型训练

本节将介绍作为 PyTorch 1.12 版本的一部分的完全分片数据并行（FSDP）的高级功能。下面以文本摘要的 FSDP 为工作示例，对 HuggingFace(HF)T5 模型进行微调。

FSDP 是一个生产就绪的软件包，专注于易用性、性能和长期支持。FSDP 的主要好处之一是减少了每个 GPU 上的内存占用。这使得能够用比 DDP 更低的总内存来训练更大的模型，并利用计算和通信的重叠来高效地训练模型。这种降低的内存压力可以用来训练更大的模型或增加批量大小，这有助于整体训练吞吐量。

FSDP 的工作原理如下：

（1）在构造函数中：

- 分片模型参数，每个等级只保留自己的分片。

（2）在前向传播中：

- 运行 all_gather 从所有等级中收集所有分片，以恢复该 FSDP 单元的完整参数向前运行计算。
- 丢弃刚刚收集的非自有参数分片以释放内存。

（3）在向后传播中：

- 运行 all_gather 从所有等级中收集所有分片，以恢复该 FSDP 单元中的完整参数运行反向计算。
- 丢弃非拥有的参数以释放内存。
- 运行 reduce_scatter 以同步梯度。

HF T5 预训练模型有 4 种不同的尺寸，从 6000 万参数的小型到 110 亿参数的 XXL。在本节中，演示使用 WikiHow 数据集对带有 FSDP 的 T5 3B 进行文本摘要的微调。本节的重点是强调 FSDP 中的不同可用功能，这些功能有助于训练 3B 参数以上的大型模型。此外，还将介绍基于 transformer 的模型的特定功能。

接下来安装 PyTorch 天天版，因为天天版中提供了一些功能，如激活检查点。

```
pip3 install --pre torch torchvision torchaudio -f https://download.pytorch.org/whl/nightly/cu113/torch_nightly.html
```

创建一个数据文件夹，从 wikihowAll.csv 和 wikihowSep.cs 下载 WikiHow 数据集，并将它们放在数据文件夹中。我们将使用 summaryzation_dataset 中的 wikihow 数据集。

接下来，将以下代码片段添加到 Python 脚本 "T5_training.py" 中。

导入必要的包，代码如下：

```
import os
import argparse
import torch
import torch.nn as nn
```

```python
import torch.nn.functional as F
import torch.optim as optim
from transformers import AutoTokenizer, GPT2TokenizerFast
from transformers import T5Tokenizer, T5ForConditionalGeneration
import functools
from torch.optim.lr_scheduler import StepLR
import torch.nn.functional as F
import torch.distributed as dist
import torch.multiprocessing as mp
from torch.nn.parallel import DistributedDataParallel as DDP
from torch.utils.data.distributed import DistributedSampler
from transformers.models.t5.modeling_t5 import T5Block

from torch.distributed.algorithms._checkpoint.checkpoint_wrapper import (
    checkpoint_wrapper,
    CheckpointImpl,
    apply_activation_checkpointing_wrapper)

from torch.distributed.fsdp import (
    FullyShardedDataParallel as FSDP,
    MixedPrecision,
    BackwardPrefetch,
    ShardingStrategy,
    FullStateDictConfig,
    StateDictType,
)
from torch.distributed.fsdp.wrap import (
    transformer_auto_wrap_policy,
    enable_wrap,
    wrap,
)
from functools import partial
from torch.utils.data import DataLoader
from pathlib import Path
from summarization_dataset import *
from transformers.models.t5.modeling_t5 import T5Block
from typing import Type
import time
import tqdm
from datetime import datetime
```

在这里，使用两个助手函数来初始化分布式训练的过程，然后在训练完成后进行清理。本节将使用 torchrun，它将自动设置工作者的 RANK 和 WORLD_SIZE。

```python
def setup():
```

```python
    # initialize the process group
    dist.init_process_group("nccl")

def cleanup():
    dist.destroy_process_group()
```

设置 HuggingFace T5 模型，代码如下：

```python
def setup_model(model_name):
    model = T5ForConditionalGeneration.from_pretrained(model_name)
    tokenizer =  T5Tokenizer.from_pretrained(model_name)
    return model, tokenizer
```

添加几个辅助函数，用于日期和格式化内存度量，代码如下：

```python
def get_date_of_run():
    """create date and time for file save uniqueness
    example: 2022-05-07-08:31:12_PM'
    """
    date_of_run = datetime.now().strftime("%Y-%m-%d-%I:%M:%S_%p")
    print(f"--> current date and time of run = {date_of_run}")
    return date_of_run

def format_metrics_to_gb(item):
    """quick function to format numbers to gigabyte and round to 4 digit precision"""
    metric_num = item / g_gigabyte
    metric_num = round(metric_num, ndigits=4)
    return metric_num
```

定义训练函数，代码如下：

```python
def train(args, model, rank, world_size, train_loader, optimizer, epoch, sampler=None):
    model.train()
    local_rank = int(os.environ['LOCAL_RANK'])
    fsdp_loss = torch.zeros(2).to(local_rank)

    if sampler:
        sampler.set_epoch(epoch)
    if rank==0:
        inner_pbar = tqdm.tqdm(
            range(len(train_loader)), colour="blue", desc="r0 Training Epoch"
        )
    for batch in train_loader:
        for key in batch.keys():
            batch[key] = batch[key].to(local_rank)
        optimizer.zero_grad()
```

```
            output = model(input_ids=batch["source_ids"],attention_
mask=batch["source_mask"],labels=batch["target_ids"] )
            loss = output["loss"]
            loss.backward()
            optimizer.step()
            fsdp_loss[0] += loss.item()
            fsdp_loss[1] += len(batch)
            if rank==0:
                inner_pbar.update(1)

    dist.all_reduce(fsdp_loss, op=dist.ReduceOp.SUM)
    train_accuracy = fsdp_loss[0] / fsdp_loss[1]


    if rank == 0:
        inner_pbar.close()
        print(
                f"Train Epoch: \t{epoch}, Loss: \t{train_accuracy:.4f}"
            )
    return train_accuracy
```

定义验证函数，代码如下：

```
def validation(model, rank, world_size, val_loader):
    model.eval()
    correct = 0
    local_rank = int(os.environ['LOCAL_RANK'])
    fsdp_loss = torch.zeros(3).to(local_rank)
    if rank == 0:
        inner_pbar = tqdm.tqdm(
            range(len(val_loader)), colour="green", desc="Validation Epoch"
        )
    with torch.no_grad():
        for batch in val_loader:
            for key in batch.keys():
                batch[key] = batch[key].to(local_rank)
                output = model(input_ids=batch["source_ids"],attention_
mask=batch["source_mask"],labels=batch["target_ids"])
            fsdp_loss[0] += output["loss"].item()  # sum up batch loss
            fsdp_loss[1] += len(batch)

            if rank==0:
                inner_pbar.update(1)

    dist.all_reduce(fsdp_loss, op=dist.ReduceOp.SUM)
    val_loss = fsdp_loss[0] / fsdp_loss[1]
```

```
        if rank == 0:
            inner_pbar.close()
            print(f"Validation Loss: {val_loss:.4f}")
        return val_loss
```

定义将模型封装在 FSDP 中的分布式训练函数，代码如下：

```
def fsdp_main(args):

    model, tokenizer = setup_model("t5-base")

    local_rank = int(os.environ['LOCAL_RANK'])
    rank = int(os.environ['RANK'])
    world_size = int(os.environ['WORLD_SIZE'])


    dataset = load_dataset('wikihow', 'all', data_dir='data/')
    print(dataset.keys())
    print("Size of train dataset: ", dataset['train'].shape)
    print("Size of Validation dataset: ", dataset['validation'].shape)


    #wikihow(tokenizer, type_path, num_samples, input_length, output_length, print_text=False)
    train_dataset = wikihow(tokenizer, 'train', 1500, 512, 150, False)
    val_dataset = wikihow(tokenizer, 'validation', 300, 512, 150, False)

    sampler1 = DistributedSampler(train_dataset, rank=rank, num_replicas=world_size, shuffle=True)
    sampler2 = DistributedSampler(val_dataset, rank=rank, num_replicas=world_size)

    setup()


    train_kwargs = {'batch_size': args.batch_size, 'sampler': sampler1}
    test_kwargs = {'batch_size': args.test_batch_size, 'sampler': sampler2}
    cuda_kwargs = {'num_workers': 2,
                   'pin_memory': True,
                   'shuffle': False}
    train_kwargs.update(cuda_kwargs)
    test_kwargs.update(cuda_kwargs)

    train_loader = torch.utils.data.DataLoader(train_dataset,**train_kwargs)
    val_loader = torch.utils.data.DataLoader(val_dataset, **test_kwargs)

    t5_auto_wrap_policy = functools.partial(
```

```python
        transformer_auto_wrap_policy,
        transformer_layer_cls={
            T5Block,
        },
    )
     sharding_strategy: ShardingStrategy = ShardingStrategy.SHARD_GRAD_OP #for Zero2 and FULL_SHARD for Zero3
    torch.cuda.set_device(local_rank)


    #init_start_event = torch.cuda.Event(enable_timing=True)
    #init_end_event = torch.cuda.Event(enable_timing=True)

    #init_start_event.record()

    bf16_ready = (
    torch.version.cuda
    and torch.cuda.is_bf16_supported()
    and LooseVersion(torch.version.cuda) >= "11.0"
    and dist.is_nccl_available()
    and nccl.version() >= (2, 10)
    )

    if bf16_ready:
        mp_policy = bfSixteen
    else:
        mp_policy = None # defaults to fp32

    # model is on CPU before input to FSDP
    model = FSDP(model,
        auto_wrap_policy=t5_auto_wrap_policy,
        mixed_precision=mp_policy,
        #sharding_strategy=sharding_strategy,
        device_id=torch.cuda.current_device())

    optimizer = optim.AdamW(model.parameters(), lr=args.lr)

    scheduler = StepLR(optimizer, step_size=1, gamma=args.gamma)
    best_val_loss = float("inf")
    curr_val_loss = float("inf")
    file_save_name = "T5-model-"

    if rank == 0:
        time_of_run = get_date_of_run()
        dur = []
```

```python
            train_acc_tracking = []
            val_acc_tracking = []
            training_start_time = time.time()

        if rank == 0 and args.track_memory:
            mem_alloc_tracker = []
            mem_reserved_tracker = []

        for epoch in range(1, args.epochs + 1):
            t0 = time.time()
            train_accuracy = train(args, model, rank, world_size, train_loader, optimizer, epoch, sampler=sampler1)
            if args.run_validation:
                curr_val_loss = validation(model, rank, world_size, val_loader)
            scheduler.step()

            if rank == 0:

                print(f"--> epoch {epoch} completed...entering save and stats zone")

                dur.append(time.time() - t0)
                train_acc_tracking.append(train_accuracy.item())

                if args.run_validation:
                    val_acc_tracking.append(curr_val_loss.item())

                if args.track_memory:
                    mem_alloc_tracker.append(
                        format_metrics_to_gb(torch.cuda.memory_allocated())
                    )
                    mem_reserved_tracker.append(
                        format_metrics_to_gb(torch.cuda.memory_reserved())
                    )
                print(f"completed save and stats zone...")

            if args.save_model and curr_val_loss < best_val_loss:

                # save
                if rank == 0:
                    print(f"--> entering save model state")

                save_policy = FullStateDictConfig(offload_to_cpu=True, rank0_only=True)
                with FSDP.state_dict_type(
                    model, StateDictType.FULL_STATE_DICT, save_policy
                ):
```

```python
                cpu_state = model.state_dict()
            #print(f"saving process: rank {rank}  done w state_dict")


            if rank == 0:
                print(f"--> saving model ...")
                currEpoch = (
                    "-" + str(epoch) + "-" + str(round(curr_val_loss.item(), 4)) + ".pt"
                )
                print(f"--> attempting to save model prefix {currEpoch}")
                save_name = file_save_name + "-" + time_of_run + "-" + currEpoch
                print(f"--> saving as model name {save_name}")

                torch.save(cpu_state, save_name)

    if curr_val_loss < best_val_loss:

        best_val_loss = curr_val_loss
        if rank==0:
            print(f"-->>>> New Val Loss Record: {best_val_loss}")

dist.barrier()
cleanup()
```

解析参数并设置主函数，代码如下：

```python
if __name__ == '__main__':
    # Training settings
    parser = argparse.ArgumentParser(description='PyTorch T5 FSDP Example')
    parser.add_argument('--batch-size', type=int, default=4, metavar='N',
                        help='input batch size for training (default: 64)')
    parser.add_argument('--test-batch-size', type=int, default=4, metavar='N',
                        help='input batch size for testing (default: 1000)')
    parser.add_argument('--epochs', type=int, default=2, metavar='N',
                        help='number of epochs to train (default: 3)')
    parser.add_argument('--lr', type=float, default=.002, metavar='LR',
                        help='learning rate (default: .002)')
    parser.add_argument('--gamma', type=float, default=0.7, metavar='M',
                        help='Learning rate step gamma (default: 0.7)')
    parser.add_argument('--no-cuda', action='store_true', default=False,
                        help='disables CUDA training')
    parser.add_argument('--seed', type=int, default=1, metavar='S',
                        help='random seed (default: 1)')
    parser.add_argument('--track_memory', action='store_false', default=True,
                        help='track the gpu memory')
```

```
    parser.add_argument('--run_validation', action='store_false', default=True,
                        help='running the validation')
    parser.add_argument('--save-model', action='store_false', default=True,
                        help='For Saving the current Model')
    args = parser.parse_args()

    torch.manual_seed(args.seed)

    fsdp_main(args)
```

使用 torchrun 进行训练，代码如下：

```
torchrun --nnodes 1 --nproc_per_node 4  T5_training.py
```

auto_wrap_policy 是 FSDP 的一个功能，它可以轻松地自动分割给定的模型，并将模型、优化器和梯度分片放入不同的 FSDP 单元中。

auto_wrap_policy 可以按如下方式创建，其中 T5Block 表示 T5 transformer 层类别。

```
t5_auto_wrap_policy = functools.partial(
    transformer_auto_wrap_policy,
    transformer_layer_cls={
        T5Block,
    },
)
torch.cuda.set_device(local_rank)


model = FSDP(model,
    fsdp_auto_wrap_policy=t5_auto_wrap_policy)
```

要想查看包装好的模型，可以轻松地打印模型，并直观地检查分片和 FSDP 单元。

FSDP 支持灵活的混合精度训练，允许任意降低精度类型（如 fp16 或 bfloat16）。目前，bfloat16 仅在 Ampere GPU 上可用，因此，需要在使用它之前确认本机支持。例如，在 V100s 上，bfloat16 仍然可以运行，但由于它是非本地运行的，所以，可能会导致速度明显减慢。

要检查 bfloat16 是否是本机支持的，可以使用以下方法：

```
bf16_ready = (
    torch.version.cuda
    and torch.cuda.is_bf16_supported()
    and LooseVersion(torch.version.cuda) >= "11.0"
    and dist.is_nccl_available()
    and nccl.version() >= (2, 10)
)
```

FSDP 中混合精度的优点之一是对参数、梯度和缓冲区的不同精度级别提供细粒度控制，代码如下：

```
fpSixteen = MixedPrecision(
    param_dtype=torch.float16,
    # Gradient communication precision.
    reduce_dtype=torch.float16,
    # Buffer precision.
    buffer_dtype=torch.float16,
)

bfSixteen = MixedPrecision(
    param_dtype=torch.bfloat16,
    # Gradient communication precision.
    reduce_dtype=torch.bfloat16,
    # Buffer precision.
    buffer_dtype=torch.bfloat16,
)

fp32_policy = MixedPrecision(
    param_dtype=torch.float32,
    # Gradient communication precision.
    reduce_dtype=torch.float32,
    # Buffer precision.
    buffer_dtype=torch.float32,
)
```

这种灵活性允许用户进行细粒度控制，例如，仅设置梯度通信以降低精度，并且所有参数/缓冲区计算都以全精度完成。这在节点内通信是主要瓶颈并且参数/缓冲区必须完全精确以避免准确性问题的情况下可能有用。可以通过以下策略来实现：

```
grad_bf16 = MixedPrecision(reduce_dtype=torch.bfloat16)
```

将相关的混合精度策略添加到 FSDP 包装器中，代码如下：

```
model = FSDP(model,
        auto_wrap_policy=t5_auto_wrap_policy,
        mixed_precision=bfSixteen)
```

在实验中，观察到通过使用 bfloat16 进行训练可以提高 4 倍的速度，并且在一些可以用于增加批量的实验中可以减少大约 30% 的内存。

FSDP 支持 device_id 参数，该参数用于初始化由 device_id 给定的设备上的输入 CPU 模块。当整个模型不适合单个 GPU，但适合主机的 CPU 内存时，这很有用。当指定 device_id 时，FSDP 将在每个 FSDP 单元的基础上将模型移动到指定的设备，避免 GPU OOM 问题，同时初始化速度比基于 CPU 的初始化快几倍。

```
torch.cuda.set_device(local_rank)

model = FSDP(model,
        auto_wrap_policy=t5_auto_wrap_policy,
        mixed_precision=bfSixteen,
        device_id=torch.cuda.current_device())
```

FSDP 分片策略默认设置为完全分片，模型参数、梯度和优化器状态在所有级别中进行分片。这种分片策略也称为 Zero3 分片。如果对 Zero2 分片策略感兴趣，其中只有优化器状态和梯度被分片，则 FSDP 通过使用"ShardingStrategy.SHARD_GRAD_OP"而不是"ShardingStrategy.FULL_SHARD"将分片策略传递给 FSDP 初始化来支持此功能，代码如下：

```
torch.cuda.set_device(local_rank)

model = FSDP(model,
        auto_wrap_policy=t5_auto_wrap_policy,
        mixed_precision=bfSixteen,
        device_id=torch.cuda.current_device(),
        sharding_strategy=ShardingStrategy.SHARD_GRAD_OP # ZERO2)
```

这将减少 FSDP 中的通信开销，在这种情况下，模型在前向和后向传递后保持完整的参数。

反向预取设置控制何时应请求下一个 FSDP 单元的参数。通过将其设置为 BACKWARD_PRE，可以开始请求下一个 FSDP 的单元参数，并在当前单元的计算开始之前更早到达。这与 all_gather 通信和梯度计算重叠，后者可以以稍高的内存消耗换取提高训练速度。可以在 FSDP 包装器中使用反向预取设置，代码如下：

```
torch.cuda.set_device(local_rank)

model = FSDP(model,
        auto_wrap_policy=t5_auto_wrap_policy,
        mixed_precision=bfSixteen,
        device_id=torch.cuda.current_device(),
        backward_prefetch = BackwardPrefetch.BACKWARD_PRE)
```

backward_prefetch 有两种模式：BACKWARD_PRE 和 BACKWARD_POST。BACKWARD_POST 意味着在当前 FSDP 单元处理完成之前，不会请求下一个 FSDP 单元的参数，从而最大限度地减少内存开销。在某些情况下，使用 BACKWARD_PRE 可以将模型训练速度提高 2%～10%，对于较大的模型，速度会有更高的改进。

为了使用 FULL_STATE_DICT 保存模型检查点（以与本地模型相同的方式保存模型），PyTorch 1.12 提供了一些实用程序来支持保存较大的模型。

首先，可以指定 FullStateDictConfig，从而只允许在等级 0 上填充 state_dict 并将其卸载到 CPU。使用此配置时，FSDP 将仅在等级 0 上收集所有模型参数，并将它们逐一卸载到 CPU。当 state_dict 最终保存时，它将仅在等级 0 上填充，并包含 CPU 张量。这避免了大于单个 GPU 内存的模型的潜在 OOM。

此功能可以按如下方式运行：

```
save_policy = FullStateDictConfig(offload_to_cpu=True, rank0_only=True)
with FSDP.state_dict_type(
        model, StateDictType.FULL_STATE_DICT, save_policy
    ):
        cpu_state = model.state_dict()
if rank == 0:
 save_name = file_save_name + "-" + time_of_run + "-" + currEpoch
 torch.save(cpu_state, save_name)
```

5.8 分布式 RPC 框架入门

本节使用两个简单的示例来演示如何使用 torch.distributed.rpc 包构建分布式训练，该包最初是作为 PyTorch v1.4 中的一个实验功能引入的。

以前的小结"分布式数据并行入门"和"用 PyTorch 编写分布式应用程序"描述了 DistributedDataParallel，它支持一种特定的训练模式，在该模式中，模型在多个进程中复制，每个进程处理输入数据的一部分。有时，可能会遇到需要不同训练模式的场景。例如：

（1）在强化学习中，从环境中获取训练数据可能相对昂贵，而模型本身可能很小。在这种情况下，生成多个并行运行的观察程序并共享一个代理可能会很有用。在这种情况下，代理在本地负责训练，但应用程序仍然需要库来在观察者和训练者之间发送和接收数据。

（2）模型可能太大，无法容纳在一台机器上的 GPU 中，因此，需要一个库来帮助将模型拆分到多台机器上。或者，可能正在实施一个参数服务器训练框架，其中模型参数和训练器位于不同的机器上。

torch.distributed.rpc 包可以帮助解决上述情况。在情况（1）中，RPC 和 RRef 允许将数据从一个工作者发送到另一个工作者，同时可以轻松地引用远程数据对象。在情况（2）中，分布式 Autograd 和分布式优化器使执行反向传递和优化器步骤就像执行本地训练一样。接下来将使用强化学习示例和语言模型示例来演示 torch.distributed.rpc 的 API。注意，本节并不旨在构建最准确或最有效的模型来解决给定的问题，相反，这里的主要目标是展示如何使用 torch.distributed.rpc 包来构建分布式训练应用程序。

5.8.1 使用 RPC 和 RRef 的分布式强化学习

本节将介绍使用 RPC 构建玩具分布式强化学习模型的步骤，以解决 OpenAI Gym 中的 CartPole-v1 问题。策略代码主要借鉴了现有的单线程示例，如下所示。我们将跳过策略设计的细节，重点讨论 RPC 的使用。

```
import torch.nn as nn
import torch.nn.functional as F

class Policy(nn.Module):

    def __init__(self):
        super(Policy, self).__init__()
        self.affine1 = nn.Linear(4, 128)
        self.dropout = nn.Dropout(p=0.6)
        self.affine2 = nn.Linear(128, 2)

    def forward(self, x):
        x = self.affine1(x)
        x = self.dropout(x)
        x = F.relu(x)
        action_scores = self.affine2(x)
        return F.softmax(action_scores, dim=1)
```

在本例中，每个观察者都创建自己的环境，并等待代理的命令来运行一个回合。在每一回合中，一个观察者最多循环 n_steps 次迭代，在每次迭代中，观察者使用 RPC 将其环境状态传递给代理并返回一个动作。然后观察者将这个动作应用到观察者的环境中，并从环境中获得奖励和下一个状态。之后，观察者使用另一个 RPC 向代理报告奖励。这显然不是最有效的观察者实现。例如，一个简单的优化可以是将当前状态和最后一次奖励打包在一个 RPC 中，以减少通信开销。然而，目标是演示 RPC API，而不是为 CartPole 构建最佳解算器。因此，在这个例子中保持逻辑简单和两个步骤的明确性。

```
import argparse
import gym
import torch.distributed.rpc as rpc

parser = argparse.ArgumentParser(
    description="RPC Reinforcement Learning Example",
    formatter_class=argparse.ArgumentDefaultsHelpFormatter,
)

parser.add_argument('--world_size', default=2, type=int, metavar='W',
                    help='number of workers')
```

```python
parser.add_argument('--log_interval', type=int, default=10, metavar='N',
                    help='interval between training status logs')
parser.add_argument('--gamma', type=float, default=0.99, metavar='G',
                    help='how much to value future rewards')
parser.add_argument('--seed', type=int, default=1, metavar='S',
                    help='random seed  for reproducibility')
args = parser.parse_args()

class Observer:

    def __init__(self):
        self.id = rpc.get_worker_info().id
        self.env = gym.make('CartPole-v1')
        self.env.seed(args.seed)

    def run_episode(self, agent_rref):
        state, ep_reward = self.env.reset(), 0
        for _ in range(10000):
            # send the state to the agent to get an action
            action = agent_rref.rpc_sync().select_action(self.id, state)

            # apply the action to the environment, and get the reward
            state, reward, done, _ = self.env.step(action)

            # report the reward to the agent for training purpose
            agent_rref.rpc_sync().report_reward(self.id, reward)

            # finishes after the number of self.env._max_episode_steps
            if done:
                break
```

代理的代码稍微复杂一些,将把它分解为多个部分。在这个例子中,代理既是训练者又是主人,因此,代理向多个分布式观察者发送命令以运行回合,并且它还在本地记录所有动作和奖励,这些动作和奖励将在每个回合之后的训练阶段使用。下面的代码显示了代理的构造函数,其中大多数行都在初始化各种组件。最后的循环在其他工作者上远程初始化观察者,并在本地保存这些观察者的 RRef。代理稍后将使用这些观察者 RRef 来发送命令。应用程序不需要担心 RRef 的使用寿命。每个 RRef 的所有者都维护一个引用计数映射来跟踪其生存期,并保证只要存在该 RRef 的任何活动用户,就不会删除远程数据对象。

```python
import gym
import numpy as np
```

```python
import torch
import torch.distributed.rpc as rpc
import torch.optim as optim
from torch.distributed.rpc import RRef, rpc_async, remote
from torch.distributions import Categorical

class Agent:
    def __init__(self, world_size):
        self.ob_rrefs = []
        self.agent_rref = RRef(self)
        self.rewards = {}
        self.saved_log_probs = {}
        self.policy = Policy()
        self.optimizer = optim.Adam(self.policy.parameters(), lr=1e-2)
        self.eps = np.finfo(np.float32).eps.item()
        self.running_reward = 0
        self.reward_threshold = gym.make('CartPole-v1').spec.reward_threshold
        for ob_rank in range(1, world_size):
            ob_info = rpc.get_worker_info(OBSERVER_NAME.format(ob_rank))
            self.ob_rrefs.append(remote(ob_info, Observer))
            self.rewards[ob_info.id] = []
            self.saved_log_probs[ob_info.id] = []
```

接下来，代理向观察者公开两个 API，用于选择动作和报告奖励。这些函数只在代理上本地运行，但将由观察者通过 RPC 触发。

```python
class Agent:
    ...
    def select_action(self, ob_id, state):
        state = torch.from_numpy(state).float().unsqueeze(0)
        probs = self.policy(state)
        m = Categorical(probs)
        action = m.sample()
        self.saved_log_probs[ob_id].append(m.log_prob(action))
        return action.item()

    def report_reward(self, ob_id, reward):
        self.rewards[ob_id].append(reward)
```

在代理上添加一个 run_epiode() 函数，该函数告诉所有观察者执行一个回合。run_epiode() 函数首先创建一个列表来收集异步 RPC 的 future，然后在所有观察者 RRef 上循环以生成异步 RPC。在这些 RPC 中，代理还将自身的 RRef 传递给观察者，以便观察者也可以调用代理上的函数。每个观察者都会将 RPC 返回到代理，这是嵌套的 RPC。在每一回合之后，saved_log_probs 和奖励将包含记录的动作问题和奖励。

```python
class Agent:
    ...
    def run_episode(self):
        futs = []
        for ob_rref in self.ob_rrefs:
            # make async RPC to kick off an episode on all observers
            futs.append(
                rpc_async(
                    ob_rref.owner(),
                    ob_rref.rpc_sync().run_episode,
                    args=(self.agent_rref,)
                )
            )

        # wait until all obervers have finished this episode
        for fut in futs:
            fut.wait()
```

最后，在一个回合之后，代理需要训练模型，这在下面的 finish_epinode() 函数中实现。

```python
class Agent:
    ...
    def finish_episode(self):
        # joins probs and rewards from different observers into lists
        R, probs, rewards = 0, [], []
        for ob_id in self.rewards:
            probs.extend(self.saved_log_probs[ob_id])
            rewards.extend(self.rewards[ob_id])

        # use the minimum observer reward to calculate the running reward
        min_reward = min([sum(self.rewards[ob_id]) for ob_id in self.rewards])
        self.running_reward = 0.05 * min_reward + (1 - 0.05) * self.running_reward

        # clear saved probs and rewards
        for ob_id in self.rewards:
            self.rewards[ob_id] = []
            self.saved_log_probs[ob_id] = []

        policy_loss, returns = [], []
        for r in rewards[::-1]:
            R = r + args.gamma * R
            returns.insert(0, R)
        returns = torch.tensor(returns)
        returns = (returns - returns.mean()) / (returns.std() + self.eps)
```

```python
        for log_prob, R in zip(probs, returns):
            policy_loss.append(-log_prob * R)
    self.optimizer.zero_grad()
    policy_loss = torch.cat(policy_loss).sum()
    policy_loss.backward()
    self.optimizer.step()
    return min_reward
```

使用 Policy、Observer 和 Agent 类，可以启动多个流程来执行分布式训练。在本例中，所有进程都运行相同的 run_worker() 函数，它们使用等级来区分自己的角色。等级 0 始终是代理，所有其他等级都是观察者。代理通过反复调用 run_epinode() 和 finish_epiode() 充当 master，直到运行的奖励超过环境指定的奖励阈值。所有观察者都被动地等待来自代理的命令。代码由 rpc.int_rpc() 和 rpc.shutdown() 包装，它们分别初始化和终止 rpc 实例。

```python
import os
from itertools import count

import torch.multiprocessing as mp

AGENT_NAME = "agent"
OBSERVER_NAME="obs{}"

def run_worker(rank, world_size):
    os.environ['MASTER_ADDR'] = 'localhost'
    os.environ['MASTER_PORT'] = '29500'
    if rank == 0:
        # rank0 is the agent
        rpc.init_rpc(AGENT_NAME, rank=rank, world_size=world_size)

        agent = Agent(world_size)
        print(f"This will run until reward threshold of {agent.reward_threshold}"
              " is reached. Ctrl+C to exit.")
        for i_episode in count(1):
            agent.run_episode()
            last_reward = agent.finish_episode()

            if i_episode % args.log_interval == 0:
                print(f"Episode {i_episode}\tLast reward: {last_reward:.2f}\tAverage reward: "
                      f"{agent.running_reward:.2f}")
            if agent.running_reward > agent.reward_threshold:
                print(f"Solved! Running reward is now {agent.running_reward}!")
                break
```

```
    else:
        # other ranks are the observer
        rpc.init_rpc(OBSERVER_NAME.format(rank), rank=rank, world_size=world_size)
        # observers passively waiting for instructions from the agent

    # block until all rpcs finish, and shutdown the RPC instance
    rpc.shutdown()


mp.spawn(
    run_worker,
    args=(args.world_size, ),
    nprocs=args.world_size,
    join=True
)
```

以下是使用 world_size=2 进行训练时的一些示例输出。

```
This will run until reward threshold of 475.0 is reached. Ctrl+C to exit.
Episode 10      Last reward: 26.00      Average reward: 10.01
Episode 20      Last reward: 16.00      Average reward: 11.27
Episode 30      Last reward: 49.00      Average reward: 18.62
Episode 40      Last reward: 45.00      Average reward: 26.09
Episode 50      Last reward: 44.00      Average reward: 30.03
Episode 60      Last reward: 111.00     Average reward: 42.23
Episode 70      Last reward: 131.00     Average reward: 70.11
Episode 80      Last reward: 87.00      Average reward: 76.51
Episode 90      Last reward: 86.00      Average reward: 95.93
Episode 100     Last reward: 13.00      Average reward: 123.93
Episode 110     Last reward: 33.00      Average reward: 91.39
Episode 120     Last reward: 73.00      Average reward: 76.38
Episode 130     Last reward: 137.00     Average reward: 88.08
Episode 140     Last reward: 89.00      Average reward: 104.96
Episode 150     Last reward: 97.00      Average reward: 98.74
Episode 160     Last reward: 150.00     Average reward: 100.87
Episode 170     Last reward: 126.00     Average reward: 104.38
Episode 180     Last reward: 500.00     Average reward: 213.74
Episode 190     Last reward: 322.00     Average reward: 300.22
Episode 200     Last reward: 165.00     Average reward: 272.71
Episode 210     Last reward: 168.00     Average reward: 233.11
Episode 220     Last reward: 184.00     Average reward: 195.02
Episode 230     Last reward: 284.00     Average reward: 208.32
Episode 240     Last reward: 395.00     Average reward: 247.37
Episode 250     Last reward: 500.00     Average reward: 335.42
Episode 260     Last reward: 500.00     Average reward: 386.30
```

```
Episode 270      Last reward: 500.00      Average reward: 405.29
Episode 280      Last reward: 500.00      Average reward: 443.29
Episode 290      Last reward: 500.00      Average reward: 464.65
Solved! Running reward is now 475.3163778435275!
```

在这个例子中，展示了如何使用 RPC 作为通信工具在工作者之间传递数据，以及如何使用 RRef 来引用远程对象。可以直接在 ProcessGroup send 和 recv API 之上构建整个结构，或者使用其他通信 /RPC 库。通过使用 torch.distributed.rpc，用户可以在后台获得本机支持和持续优化的性能。

接下来，将展示如何将 RPC 和 RRef 与分布式 Autograd 和分布式优化器相结合，以执行分布式模型并行训练。

5.8.2 使用分布式 Autograd 和分布式优化器的分布式 RNN

本节使用 RNN 模型来展示如何使用 RPC API 构建分布式模型并行训练。示例 RNN 模型非常小，可以很容易地放入一个 GPU 中，但我们仍然将其划分为两个不同的工作者来演示这个想法。开发人员可以应用类似的技术在多个设备和机器上分布更大的模型。

RNN 模型设计借鉴了 PyTorch 示例库中的单词语言模型，该模型包含 3 个主要组件：1 个嵌入表、1 个 LSTM 层和 1 个解码器。下面的代码将嵌入表和解码器包装成子模块，以便将它们的构造函数传递给 RPC API。在 EmbeddingTable 子模块中，将嵌入层放在 GPU 上以覆盖用例。RPC 总是在目标工作者上创建 CPU 张量参数或返回值。如果函数采用 GPU 张量，则需要明确地将其移动到适当的设备中。

```
class EmbeddingTable(nn.Module):
    r"""
    Encoding layers of the RNNModel
    """
    def __init__(self, ntoken, ninp, dropout):
        super(EmbeddingTable, self).__init__()
        self.drop = nn.Dropout(dropout)
        self.encoder = nn.Embedding(ntoken, ninp).cuda()
        self.encoder.weight.data.uniform_(-0.1, 0.1)

    def forward(self, input):
        return self.drop(self.encoder(input.cuda())).cpu()


class Decoder(nn.Module):
    def __init__(self, ntoken, nhid, dropout):
        super(Decoder, self).__init__()
        self.drop = nn.Dropout(dropout)
```

```
            self.decoder = nn.Linear(nhid, ntoken)
            self.decoder.bias.data.zero_()
            self.decoder.weight.data.uniform_(-0.1, 0.1)

        def forward(self, output):
            return self.decoder(self.drop(output))
```

有了上面的子模块，现在可以使用 RPC 将它们拼凑在一起，以创建 RNN 模型。在下面的代码中，ps 代表一个参数服务器，它托管嵌入表和解码器的参数。构造函数使用远程 API 在参数服务器上创建 EmbeddingTable 对象和 Decoder 对象，并在本地创建 LSTM 子模块。在前向传播过程中，训练器使用 EmbeddingTable RRef 查找远程子模块，并使用 RPC 将输入数据传递给 EmbeddingTable 并获取查找结果。然后，参数服务器通过本地 LSTM 层运行嵌入，最后使用另一个 RPC 将输出发送到解码器子模块。通常，为了实现分布式模型并行训练，开发人员可以将模型划分为子模块，调用 RPC 远程创建子模块实例，并在必要时使用 RRef 来查找它们。分布式模型并行训练看起来非常类似于单机模型并行训练，主要区别在于用 RPC 函数替换 Tensor.to(device)。

```
class RNNModel(nn.Module):
    def __init__(self, ps, ntoken, ninp, nhid, nlayers, dropout=0.5):
        super(RNNModel, self).__init__()

        # setup embedding table remotely
        self.emb_table_rref = rpc.remote(ps, EmbeddingTable, args=(ntoken, ninp, dropout))
        # setup LSTM locally
        self.rnn = nn.LSTM(ninp, nhid, nlayers, dropout=dropout)
        # setup decoder remotely
        self.decoder_rref = rpc.remote(ps, Decoder, args=(ntoken, nhid, dropout))

    def forward(self, input, hidden):
        # pass input to the remote embedding table and fetch emb tensor back
        emb = _remote_method(EmbeddingTable.forward, self.emb_table_rref, input)
        output, hidden = self.rnn(emb, hidden)
        # pass output to the rremote decoder and get the decoded output back
        decoded = _remote_method(Decoder.forward, self.decoder_rref, output)
        return decoded, hidden
```

在介绍分布式优化器之前，先添加一个助手函数来生成模型参数的 RRef 列表，这些参数将由分布式优化器使用。在本地训练中，应用程序可以调用 Module.parameters() 来获取对所有参数张量的引用，并将其传递给本地优化器以进行后续更新。然而，相同的 API 在分布式训练场景中不起作用，因为一些参数存在于远程机器上，所以，分布式优化器不采用参数张量列表，而是采用 RRef 列表，即本地模型参数和远程模型参数的每个模型参数一个 RRef。helper 函数非常简单，只需调用 Module.parameters() 并在每个参数上创建一个本地 RRef 即可。

```
def _parameter_rrefs(module):
    param_rrefs = []
    for param in module.parameters():
        param_rrefs.append(RRef(param))
    return param_rrefs
```

然后，因为 RNNModel 包含 3 个子模块，所以，需要调用 _parameter_rrefs 3 次，并将其封装到另一个辅助函数中。

```
class RNNModel(nn.Module):
    ...
    def parameter_rrefs(self):
        remote_params = []
        # get RRefs of embedding table
        remote_params.extend(_remote_method(_parameter_rrefs, self.emb_table_rref))
        # create RRefs for local parameters
        remote_params.extend(_parameter_rrefs(self.rnn))
        # get RRefs of decoder
        remote_params.extend(_remote_method(_parameter_rrefs, self.decoder_rref))
        return remote_params
```

现在，已经准备好执行训练循环。初始化模型参数后，创建 RNNModel 和 DistributedOptimizer。分布式优化器将获取一个参数 RRef 列表，找到所有不同的所有者工作者，并使用给定的参数（即 lr=0.05）在每个所有者工作者上创建给定的本地优化器。

在训练循环中，分布式优化器首先创建一个分布式 Autograd 上下文，这将帮助分布式 Autograd 引擎找到梯度和涉及的 RPC 发送/接收函数。然后，分布式优化器像本地模型一样启动正向传播，并运行分布式反向传播。对于分布式反向传播，只需要指定一个根列表，在这种情况下，根列表就是 loss 张量。分布式自动梯度引擎将自动遍历分布式图并正确写入梯度。接下来，分布式自动梯度引擎在分布式优化器上运行 step() 函数，分布式自动梯度引擎将联系所有相关的本地优化器来更新模型参数。与局部训练相比，一个小的区别是不需要运行 zero_grad()，因为每个 Autograd 上下文都有专门的空间来存储梯度，每次迭代都创建一个上下文时，来自不同迭代的梯度不会累积到同一组张量中。

```
def run_trainer():
    batch = 5
    ntoken = 10
    ninp = 2

    nhid = 3
    nindices = 3
```

```python
    nlayers = 4
    hidden = (
        torch.randn(nlayers, nindices, nhid),
        torch.randn(nlayers, nindices, nhid)
    )

    model = rnn.RNNModel('ps', ntoken, ninp, nhid, nlayers)

    # setup distributed optimizer
    opt = DistributedOptimizer(
        optim.SGD,
        model.parameter_rrefs(),
        lr=0.05,
    )

    criterion = torch.nn.CrossEntropyLoss()

    def get_next_batch():
        for _ in range(5):
            data = torch.LongTensor(batch, nindices) % ntoken
            target = torch.LongTensor(batch, ntoken) % nindices
            yield data, target

    # train for 10 iterations
    for epoch in range(10):
        for data, target in get_next_batch():
            # create distributed autograd context
            with dist_autograd.context() as context_id:
                hidden[0].detach_()
                hidden[1].detach_()
                output, hidden = model(data, hidden)
                loss = criterion(output, target)
                # run distributed backward pass
                dist_autograd.backward(context_id, [loss])
                # run distributed optimizer
                opt.step(context_id)
                # not necessary to zero grads since they are
                # accumulated into the distributed autograd context
                # which is reset every iteration.
        print("Training epoch {}".format(epoch))
```

最后，添加一些黏合代码来启动参数服务器和训练器进程。

```python
def run_worker(rank, world_size):
    os.environ['MASTER_ADDR'] = 'localhost'
```

```
        os.environ['MASTER_PORT'] = '29500'
        if rank == 1:
            rpc.init_rpc("trainer", rank=rank, world_size=world_size)
            _run_trainer()
        else:
            rpc.init_rpc("ps", rank=rank, world_size=world_size)
            # parameter server do nothing
            pass

        # block until all rpcs finish
        rpc.shutdown()


if __name__=="__main__":
    world_size = 2
    mp.spawn(run_worker, args=(world_size, ), nprocs=world_size, join=True)
```

5.9 使用分布式 RPC 框架实现参数服务器

本节将介绍一个使用 PyTorch 的分布式 RPC 框架实现参数服务器的简单示例。参数服务器框架是一个范式，其中一组服务器存储参数（如大型嵌入表），并且多个训练器查询参数服务器，以便检索最新的参数。这些训练器可以在本地运行训练循环，偶尔与参数服务器同步，以获得最新的参数。

使用分布式 RPC 框架，将构建一个示例，其中多个训练器使用 RPC 与同一参数服务器通信，并使用 RRef 访问远程参数服务器实例上的状态。每个训练器将通过使用分布式 Autograd 在多个节点上缝合 Autograd 图，以分布式方式启动其专用的后向传播。

导入所需的模块，并定义一个将在 MNIST 数据集上训练的简单 ConvNet。

```
import argparse
import os
import time
from threading import Lock

import torch
import torch.distributed.autograd as dist_autograd
import torch.distributed.rpc as rpc
import torch.multiprocessing as mp
import torch.nn as nn
import torch.nn.functional as F
from torch import optim
from torch.distributed.optim import DistributedOptimizer
```

```python
from torchvision import datasets, transforms

# --------- MNIST Network to train, from pytorch/examples -----

class Net(nn.Module):
    def __init__(self, num_gpus=0):
        super(Net, self).__init__()
        print(f"Using {num_gpus} GPUs to train")
        self.num_gpus = num_gpus
        device = torch.device(
            "cuda:0" if torch.cuda.is_available() and self.num_gpus > 0 else "cpu")
        print(f"Putting first 2 convs on {str(device)}")
        # Put conv layers on the first cuda device, or CPU if no cuda device
        self.conv1 = nn.Conv2d(1, 32, 3, 1).to(device)
        self.conv2 = nn.Conv2d(32, 64, 3, 1).to(device)
        # Put rest of the network on the 2nd cuda device, if there is one
        if "cuda" in str(device) and num_gpus > 1:
            device = torch.device("cuda:1")

        print(f"Putting rest of layers on {str(device)}")
        self.dropout1 = nn.Dropout2d(0.25).to(device)
        self.dropout2 = nn.Dropout2d(0.5).to(device)
        self.fc1 = nn.Linear(9216, 128).to(device)
        self.fc2 = nn.Linear(128, 10).to(device)

    def forward(self, x):
        x = self.conv1(x)
        x = F.relu(x)
        x = self.conv2(x)
        x = F.max_pool2d(x, 2)

        x = self.dropout1(x)
        x = torch.flatten(x, 1)
        # Move tensor to next device if necessary
        next_device = next(self.fc1.parameters()).device
        x = x.to(next_device)

        x = self.fc1(x)
        x = F.relu(x)
        x = self.dropout2(x)
        x = self.fc2(x)
        output = F.log_softmax(x, dim=1)
        return output
```

接下来，定义一些对脚本的其余部分有用的辅助函数。下面使用 rpc_sync 和 RRef 来定义一个函数，该函数调用位于远程节点上的对象上的给定方法。下面，我们对远程对象的句柄由 rref 参数给出，并且在其拥有节点 rref.owner() 上运行它。在调用方节点上，通过使用 rpc_sync 同步运行此命令，这意味着将阻塞直到收到响应。

```python
# --------- Helper Methods --------------------

# On the local node, call a method with first arg as the value held by the
# RRef. Other args are passed in as arguments to the function called.
# Useful for calling instance methods. method could be any matching function, including
# class methods.
def call_method(method, rref, *args, **kwargs):
    return method(rref.local_value(), *args, **kwargs)

# Given an RRef, return the result of calling the passed in method on the value
# held by the RRef. This call is done on the remote node that owns
# the RRef and passes along the given argument.
# Example: If the value held by the RRef is of type Foo, then
# remote_method(Foo.bar, rref, arg1, arg2) is equivalent to calling
# <foo_instance>.bar(arg1, arg2) on the remote node and getting the result
# back.

def remote_method(method, rref, *args, **kwargs):
    args = [method, rref] + list(args)
    return rpc.rpc_sync(rref.owner(), call_method, args=args, kwargs=kwargs)
```

现在，已经准备好定义参数服务器了。接下来将对 nn.Module 进行子类化，并保存上面定义的网络的句柄。我们还将保存一个输入设备，它将是调用模型之前输入传输到的设备。

```python
# --------- Parameter Server --------------------
class ParameterServer(nn.Module):
    def __init__(self, num_gpus=0):
        super().__init__()
        model = Net(num_gpus=num_gpus)
        self.model = model
        self.input_device = torch.device(
            "cuda:0" if torch.cuda.is_available() and num_gpus > 0 else "cpu")
```

接下来，将定义向前传播。无论模型输出的设备是什么，都会将输出移动到 CPU，因为分布式 RPC 框架目前只支持通过 RPC 发送 CPU 张量。由于调用者/被调用者上可能存在不同的设备（CPU/GPU），所以，禁用通过 RPC 发送 CUDA 张量，但可能在未来的版本中支持此功能。

```
class ParameterServer(nn.Module):
...
    def forward(self, inp):
        inp = inp.to(self.input_device)
        out = self.model(inp)
        # This output is forwarded over RPC, which as of 1.5.0 only accepts CPU tensors.
        # Tensors must be moved in and out of GPU memory due to this.
        out = out.to("cpu")
        return out
```

接下来,定义一些对训练和验证有用的杂项函数。第一个是 get_dist_gradients(),它将接收分布式 Autograd 上下文 ID,并调用 dist_autograd.get_gradients API,以便检索由分布式 Autograd 计算的梯度。请注意,我们还遍历生成的字典,并将每个张量转换为 CPU 张量,因为该框架目前只支持通过 RPC 发送张量。接下来,get_param_rrefs() 将遍历模型参数,并将它们包装为 RRef。get_param_rrefs() 方法将由训练器节点通过 RPC 调用,并返回要优化的参数列表。这需要作为分布式优化器的输入,分布式优化器需要分布式优化器必须优化的所有参数作为 RRef 列表。

```
# Use dist autograd to retrieve gradients accumulated for this model.
# Primarily used for verification.
def get_dist_gradients(self, cid):
    grads = dist_autograd.get_gradients(cid)
    # This output is forwarded over RPC, which as of 1.5.0 only accepts CPU tensors.
    # Tensors must be moved in and out of GPU memory due to this.
    cpu_grads = {}
    for k, v in grads.items():
        k_cpu, v_cpu = k.to("cpu"), v.to("cpu")
        cpu_grads[k_cpu] = v_cpu
    return cpu_grads

# Wrap local parameters in a RRef. Needed for building the
# DistributedOptimizer which optimizes paramters remotely.
def get_param_rrefs(self):
    param_rrefs = [rpc.RRef(param) for param in self.model.parameters()]
    return param_rrefs
```

最后,将创建初始化参数服务器的方法。在所有进程中,参数服务器只有一个实例,所有训练器都将与同一个参数服务器对话,并更新同一个存储的模型。服务器本身不采取任何独立的操作,而是等待来自训练器的请求,并通过运行请求的函数来响应它们。

```
# The global parameter server instance.
param_server = None
# A lock to ensure we only have one parameter server.
global_lock = Lock()
```

```python
def get_parameter_server(num_gpus=0):
    """
    Returns a singleton parameter server to all trainer processes
    """
    global param_server
    # Ensure that we get only one handle to the ParameterServer.
    with global_lock:
        if not param_server:
            # construct it once
            param_server = ParameterServer(num_gpus=num_gpus)
        return param_server

def run_parameter_server(rank, world_size):
    # The parameter server just acts as a host for the model and responds to
    # requests from trainers.
    # rpc.shutdown() will wait for all workers to complete by default, which
    # in this case means that the parameter server will wait for all trainers
    # to complete, and then exit.
    print("PS master initializing RPC")
    rpc.init_rpc(name="parameter_server", rank=rank, world_size=world_size)
    print("RPC initialized! Running parameter server...")
    rpc.shutdown()
    print("RPC shutdown on parameter server.")
```

rpc.shutdown() 不会立即关闭参数服务器，它将等待所有工作者（在本例中为训练器）调用 rpc.shutdown()。这保证了在所有训练器完成训练过程之前，参数服务器不会离线。

接下来将定义 TrainerNet 类。TrainerNet 类是 nn.Module 的一个子类，__init__ 方法将使用 rpc.remote API 来获取对参数服务器的 RRef 或远程引用。这里并没有将参数服务器复制到本地进程，相反，可以将 self.param_server_rref 视为指向位于单独进程上的参数服务器的分布式共享指针。

```python
# --------- Trainers --------------------

# nn.Module corresponding to the network trained by this trainer. The
# forward() method simply invokes the network on the given parameter
# server.
class TrainerNet(nn.Module):
    def __init__(self, num_gpus=0):
        super().__init__()
        self.num_gpus = num_gpus
        self.param_server_rref = rpc.remote(
```

```
"parameter_server", get_parameter_server, args=(num_gpus,))
```

接下来将定义一个名为 get_global_param_rrefs() 的方法。必须向优化器传递与要优化的远程参数相对应的 RRef 列表,因此在这里获得必要的 RRef。由于给定 TrainerNet 与之交互的唯一远程工作者是 ParameterServer,所以,只需在 ParameterServer 上调用 remote_method()。使用在 ParameterServer 类中定义的 get_param_rrefs() 方法。此方法将向需要优化的参数返回一个 RRef 列表。在这种情况下,TrainerNet 没有定义自己的参数;如果是这样,还需要将每个参数封装在 RRef 中,并将其包含在 DistributedOptimizer 的输入中。

```
class TrainerNet(nn.Module):
...
    def get_global_param_rrefs(self):
        remote_params = remote_method(
            ParameterServer.get_param_rrefs,
            self.param_server_rref)
        return remote_params
```

现在,准备定义 forward() 方法,它将调用 RPC 来运行在 ParameterServer 上定义的网络的正向传播。将 self.param_server_ref(ParameterServer 的远程句柄)传递给 RPC 调用。此调用将向运行 ParameterServer 的节点发送 RPC,调用前向传播,并返回与模型输出相对应的张量。

```
class TrainerNet(nn.Module):
...
    def forward(self, x):
        model_output = remote_method(
            ParameterServer.forward, self.param_server_rref, x)
        return model_output
```

训练器完全定义后,编写神经网络训练循环,它将创建网络和优化器,通过网络运行一些输入并计算损失。训练循环看起来很像本地训练程序,由于网络分布在机器之间,因此进行了一些修改。

下面,初始化 TrainerNet 并构建一个 DistributedOptimizer。注意,必须传入想要优化的所有全局参数。此外,还需要传入要使用的本地优化器,在本例中为 SGD。

```
def run_training_loop(rank, num_gpus, train_loader, test_loader):
    # Runs the typical nueral network forward + backward + optimizer step, but
    # in a distributed fashion.
    net = TrainerNet(num_gpus=num_gpus)
    # Build DistributedOptimizer.
    param_rrefs = net.get_global_param_rrefs()
    opt = DistributedOptimizer(optim.SGD, param_rrefs, lr=0.03)
```

接下来,定义主要训练循环。循环遍历 PyTorch 的 DataLoader 提供的可迭代项。在编写典型的前向/后向/优化器循环之前,首先将逻辑封装在分布式 Autograd 上下文中。这是记录在模型的前向传播中调用的 RPC 所必需的,这样就可以构建一个适当的图,其中包括后向传播中所有参与的分布式工作者。分布式 Autograd 上下文返回 context_id,该 context_id 用作累积和优化与特定迭代相对应的梯度的标识符。

```
def run_training_loop(rank, num_gpus, train_loader, test_loader):
    ...
    for i, (data, target) in enumerate(train_loader):
        with dist_autograd.context() as cid:
            model_output = net(data)
            target = target.to(model_output.device)
            loss = F.nll_loss(model_output, target)
            if i % 5 == 0:
                print(f"Rank {rank} training batch {i} loss {loss.item()}")
            dist_autograd.backward(cid, [loss])
            # Ensure that dist autograd ran successfully and gradients were
            # returned.
            assert remote_method(
                ParameterServer.get_dist_gradients,
                net.param_server_rref,
                cid) != {}
            opt.step(cid)

    print("Training complete!")
    print("Getting accuracy....")
    get_accuracy(test_loader, net)
```

下面计算完成训练后模型的准确性,很像传统的本地模型。但是,请注意,我们在上面传递给该函数的网络是 TrainerNet 的一个实例,因此,前向传播以透明的方式调用 RPC。

```
def get_accuracy(test_loader, model):
    model.eval()
    correct_sum = 0
    # Use GPU to evaluate if possible
    device = torch.device("cuda:0" if model.num_gpus > 0
        and torch.cuda.is_available() else "cpu")
    with torch.no_grad():
        for i, (data, target) in enumerate(test_loader):
            out = model(data, -1)
            pred = out.argmax(dim=1, keepdim=True)
            pred, target = pred.to(device), target.to(device)
            correct = pred.eq(target.view_as(pred)).sum().item()
            correct_sum += correct
```

```
print(f"Accuracy {correct_sum / len(test_loader.dataset)}")
```

接下来，类似于如何将 run_parameter_server 定义为负责初始化 RPC 的 ParameterServer 的主循环，为训练器定义一个类似的循环。不同之处在于，训练器必须运行上面定义的训练循环。

```
# Main loop for trainers.
def run_worker(rank, world_size, num_gpus, train_loader, test_loader):
    print(f"Worker rank {rank} initializing RPC")
    rpc.init_rpc(
        name=f"trainer_{rank}",
        rank=rank,
        world_size=world_size)

    print(f"Worker {rank} done initializing RPC")

    run_training_loop(rank, num_gpus, train_loader, test_loader)
    rpc.shutdown()
```

与 run_parameter_server 类似，rpc.shutdown() 默认情况下会等待所有工作者在该节点退出之前调用 rpc.shutdown()。这样可以确保节点正常终止，并且当另一个节点期望该节点处于联机状态时，没有任何节点处于脱机状态。

现在已经完成了特定于训练器和参数服务器的代码，剩下的就是添加代码来启动训练器和参数服务器。首先，必须考虑适用于参数服务器和训练器的各种参数。world_size 对应于将参与训练的节点总数，是所有训练器和参数服务器的总和。还必须为每个单独的进程传递一个唯一的等级，从 0（将在其中运行参数服务器）到 world_size-1。master_addr 和 master_port 是可用于标识等级 0 进程运行位置的参数，并且将由各个节点用于发现彼此。要在本地测试这个示例，只需将 localhost 和相同的 master_port 传递给派生的所有实例。出于演示目的，此示例仅支持 0～2 个 GPU。可以扩展该模式以使用其他 GPU。

```
if __name__ == '__main__':
    parser = argparse.ArgumentParser(
        description="Parameter-Server RPC based training")
    parser.add_argument(
        "--world_size",
        type=int,
        default=4,
        help="""Total number of participating processes. Should be the sum of
        master node and all training nodes.""")
    parser.add_argument(
        "rank",
        type=int,
```

```
            default=None,
            help="Global rank of this process. Pass in 0 for master.")
        parser.add_argument(
            "num_gpus",
            type=int,
            default=0,
            help="""Number of GPUs to use for training, Currently supports between 0
            and 2 GPUs. Note that this argument will be passed to the parameter
servers.""")
        parser.add_argument(
            "--master_addr",
            type=str,
            default="localhost",
            help="""Address of master, will default to localhost if not provided.
            Master must be able to accept network traffic on the address + port.""")
        parser.add_argument(
            "--master_port",
            type=str,
            default="29500",
            help="""Port that master is listening on, will default to 29500 if not
            provided. Master must be able to accept network traffic on the host and
port.""")

        args = parser.parse_args()
        assert args.rank is not None, "must provide rank argument."
        assert args.num_gpus <= 3, f"Only 0-2 GPUs currently supported (got {args.num_gpus})."
        os.environ['MASTER_ADDR'] = args.master_addr
        os.environ["MASTER_PORT"] = args.master_port
```

根据命令行参数创建一个与参数服务器或训练器相对应的进程。如果传入的等级为 0，将创建一个 ParameterServer；否则，将创建 TrainerNet。使用 torch.multiprocessing 启动与要执行的函数相对应的子进程，并使用 p.join() 从主线程等待该进程完成。在初始化训练器的情况下，使用 PyTorch 的数据加载器来指定 MNIST 数据集上的训练和测试数据加载器。

```
    processes = []
    world_size = args.world_size
    if args.rank == 0:
        p = mp.Process(target=run_parameter_server, args=(0, world_size))
        p.start()
        processes.append(p)
    else:
        # Get data to train on
```

```python
    train_loader = torch.utils.data.DataLoader(
        datasets.MNIST('../data', train=True, download=True,
                       transform=transforms.Compose([
                           transforms.ToTensor(),
                           transforms.Normalize((0.1307,), (0.3081,))
                       ])),
        batch_size=32, shuffle=True,)
    test_loader = torch.utils.data.DataLoader(
        datasets.MNIST(
            '../data',
            train=False,
            transform=transforms.Compose([
                transforms.ToTensor(),
                transforms.Normalize((0.1307,), (0.3081,))
                ])),
        batch_size=32,
        shuffle=True,
    )
    # start training worker on this node
    p = mp.Process(
        target=run_worker,
        args=(
            args.rank,
            world_size, args.num_gpus,
            train_loader,
            test_loader))
    p.start()
    processes.append(p)

for p in processes:
    p.join()
```

要在本地运行该示例，需在单独的终端窗口中为服务器和要生成的每个工作进程运行以下命令：python rpc_parameter_server.py --world_size=WORLD_SIZE --rank=RANK。例如，对于世界大小为 2 的主节点，命令为 python rpc_parameter_server.py --world_size=2 --rank=0。然后，可以在一个单独的窗口中使用命令 python rpc_parameter_server.py --world_size=2 --rank=1 启动训练器，这将开始使用一台服务器和一台训练器进行训练。注意，本节假设使用 0 ~ 2 个 GPU 进行训练，并且可以通过将 --num_gpus=N 传递到训练脚本中来配置此参数。可以传入命令行参数 --master_addr=ADDRESS 和 --master_port=PORT，以指示主工作进程正在侦听的地址和端口。

5.10 基于 RPC 的分布式流水线并行

本节使用 Resnet50 模型来演示使用 torch.distributed.rpc API 实现分布式管道并行性。本节将分 4 个步骤介绍实现。

5.10.1 步骤 1：分区 ResNet50 模型

这是在两个模型分片中实现 ResNet50 的准备步骤。ResNetBase 模块包含两个 ResNet 分片的公共构建块和属性。

```
import threading

import torch
import torch.nn as nn

from torchvision.models.resnet import Bottleneck

num_classes = 1000


def conv1x1(in_planes, out_planes, stride=1):
    return nn.Conv2d(in_planes, out_planes, kernel_size=1, stride=stride, bias=False)


class ResNetBase(nn.Module):
    def __init__(self, block, inplanes, num_classes=1000,
                 groups=1, width_per_group=64, norm_layer=None):
        super(ResNetBase, self).__init__()

        self._lock = threading.Lock()
        self._block = block
        self._norm_layer = nn.BatchNorm2d
        self.inplanes = inplanes
        self.dilation = 1
        self.groups = groups
        self.base_width = width_per_group

    def _make_layer(self, planes, blocks, stride=1):
        norm_layer = self._norm_layer
        downsample = None
        previous_dilation = self.dilation
        if stride != 1 or self.inplanes != planes * self._block.expansion:
            downsample = nn.Sequential(
```

```
                    conv1x1(self.inplanes, planes * self._block.expansion, stride),
                    norm_layer(planes * self._block.expansion),
                )

            layers = []
            layers.append(self._block(self.inplanes, planes, stride, downsample, self.groups,
                                  self.base_width, previous_dilation, norm_layer))
            self.inplanes = planes * self._block.expansion
            for _ in range(1, blocks):
                layers.append(self._block(self.inplanes, planes, groups=self.groups,
                                      base_width=self.base_width, dilation=self.dilation,
                                      norm_layer=norm_layer))

            return nn.Sequential(*layers)

        def parameter_rrefs(self):
            return [RRef(p) for p in self.parameters()]
```

现在，已经准备好定义这两个模型分片。对于构造函数，只需将所有ResNet50层拆分为两部分，并将每个部分移动到所提供的设备中。两个分片的forward()函数获取输入数据的RRef，forward()函数在本地获取数据，然后将其移动到预期的设备。在将所有层应用于输入之后，它将输出移动到CPU并返回。这是因为当调用者和被调用者中的设备数量不匹配时，RPC API要求张量驻留在CPU，以避免无效的设备错误。

```
class ResNetShard1(ResNetBase):
    def __init__(self, device, *args, **kwargs):
        super(ResNetShard1, self).__init__(
            Bottleneck, 64, num_classes=num_classes, *args, **kwargs)

        self.device = device
        self.seq = nn.Sequential(
                nn.Conv2d(3, self.inplanes, kernel_size=7, stride=2, padding=3, bias=False),
                self._norm_layer(self.inplanes),
                nn.ReLU(inplace=True),
                nn.MaxPool2d(kernel_size=3, stride=2, padding=1),
                self._make_layer(64, 3),
                self._make_layer(128, 4, stride=2)
        ).to(self.device)

        for m in self.modules():
            if isinstance(m, nn.Conv2d):
```

```python
                    nn.init.kaiming_normal_(m.weight, mode='fan_out',
nonlinearity='relu')
            elif isinstance(m, nn.BatchNorm2d):
                nn.init.constant_(m.weight, 1)
                nn.init.constant_(m.bias, 0)

    def forward(self, x_rref):
        x = x_rref.to_here().to(self.device)
        with self._lock:
            out = self.seq(x)
        return out.cpu()


class ResNetShard2(ResNetBase):
    def __init__(self, device, *args, **kwargs):
        super(ResNetShard2, self).__init__(
            Bottleneck, 512, num_classes=num_classes, *args, **kwargs)

        self.device = device
        self.seq = nn.Sequential(
            self._make_layer(256, 6, stride=2),
            self._make_layer(512, 3, stride=2),
            nn.AdaptiveAvgPool2d((1, 1)),
        ).to(self.device)

        self.fc = nn.Linear(512 * self.block.expansion, num_classes).to(self.device)

    def forward(self, x_rref):
        x = x_rref.to_here().to(self.device)
        with self._lock:
            out = self.fc(torch.flatten(self.seq(x), 1))
        return out.cpu()
```

5.10.2 步骤 2：将 ResNet50 模型分片拼接到一个模块中

接下来创建一个 DistResNet50 模块来组装两个分片，并实现流水线并行逻辑。在构造函数中，使用两个 rpc.remote 调用将两个分片，分别放在两个不同的 rpc 工作线程上，并保留对两个模型部分的 RRef，以便在前向传播中引用它们。forward() 函数将输入批次拆分为多个微批次，并以流水线方式将这些微批次提供给两个模型部件。首先使用 rpc.remote 调用将第一个分片应用到微批次，然后将返回的中间输出 RRef 转发到第二个模型分片。之后，收集所有微输出的 future，并在循环后等待所有这些 future。注意，remote()

和 rpc_async() 都会立即返回并异步运行。因此，整个循环是非阻塞的，并且将同时启动多个 RPC。一个微批在两个模型部件上的执行顺序由中间输出 y_rref 保留。跨微批的执行顺序无关紧要。最后，forward() 函数将所有微批的输出连接到一个单独的输出张量中并返回。parameter_rrefs() 函数是简化分布式优化器构造的助手，稍后将使用它。

```python
class DistResNet50(nn.Module):
    def __init__(self, num_split, workers, *args, **kwargs):
        super(DistResNet50, self).__init__()

        self.num_split = num_split

        # Put the first part of the ResNet50 on workers[0]
        self.p1_rref = rpc.remote(
            workers[0],
            ResNetShard1,
            args = ("cuda:0",) + args,
            kwargs = kwargs
        )

        # Put the second part of the ResNet50 on workers[1]
        self.p2_rref = rpc.remote(
            workers[1],
            ResNetShard2,
            args = ("cuda:1",) + args,
            kwargs = kwargs
        )

    def forward(self, xs):
        out_futures = []
        for x in iter(xs.split(self.num_split, dim=0)):
            x_rref = RRef(x)
            y_rref = self.p1_rref.remote().forward(x_rref)
            z_fut = self.p2_rref.rpc_async().forward(y_rref)
            out_futures.append(z_fut)

        return torch.cat(torch.futures.wait_all(out_futures))

    def parameter_rrefs(self):
        remote_params = []
        remote_params.extend(self.p1_rref.remote().parameter_rrefs().to_here())
        remote_params.extend(self.p2_rref.remote().parameter_rrefs().to_here())
        return remote_params
```

5.10.3 步骤 3：定义训练循环

在定义了模型之后，下面实现训练循环。这里使用一个专门的主工作者来准备随机输入和标签，并控制分布式反向传播和分布式优化器步骤。主工作者首先创建 DistResNet50 模块的一个实例。主工作者为每个批指定微批的数量，还提供了两个 RPC 工作程序（即 "worker1" 和 "worker2"）的名称。然后，主工作者定义损失函数，并使用 parameter_rrefs() 辅助函数创建 DistributedOptimizer，以获取参数 RRefs 的列表。然后，主训练循环与常规的本地训练非常相似，只是主训练循环使用 dist_autograd 向后启动，并为向后和优化器 step() 函数提供 context_id。

```python
import torch.distributed.autograd as dist_autograd
import torch.optim as optim
from torch.distributed.optim import DistributedOptimizer

num_batches = 3
batch_size = 120
image_w = 128
image_h = 128


def run_master(num_split):
    # put the two model parts on worker1 and worker2 respectively
    model = DistResNet50(num_split, ["worker1", "worker2"])
    loss_fn = nn.MSELoss()
    opt = DistributedOptimizer(
        optim.SGD,
        model.parameter_rrefs(),
        lr=0.05,
    )

    one_hot_indices = torch.LongTensor(batch_size) \
                        .random_(0, num_classes) \
                        .view(batch_size, 1)

    for i in range(num_batches):
        print(f"Processing batch {i}")
        # generate random inputs and labels
        inputs = torch.randn(batch_size, 3, image_w, image_h)
        labels = torch.zeros(batch_size, num_classes) \
                    .scatter_(1, one_hot_indices, 1)

        with dist_autograd.context() as context_id:
```

```
        outputs = model(inputs)
        dist_autograd.backward(context_id, [loss_fn(outputs, labels)])
        opt.step(context_id)
```

5.10.4 步骤4：启动 RPC 进程

下面的代码显示了所有进程的目标函数。主逻辑在 run_master() 中定义。工作者被动地等待来自主机的命令，因此，只需运行 init_rpc 和 shutdown，默认情况下，将阻止 shutdown，直到所有 rpc 参与者结束。

```python
import os
import time

import torch.multiprocessing as mp


def run_worker(rank, world_size, num_split):
    os.environ['MASTER_ADDR'] = 'localhost'
    os.environ['MASTER_PORT'] = '29500'
    options = rpc.TensorPipeRpcBackendOptions(num_worker_threads=128)

    if rank == 0:
        rpc.init_rpc(
            "master",
            rank=rank,
            world_size=world_size,
            rpc_backend_options=options
        )
        run_master(num_split)
    else:
        rpc.init_rpc(
            f"worker{rank}",
            rank=rank,
            world_size=world_size,
            rpc_backend_options=options
        )
        pass

    # block until all rpcs finish
    rpc.shutdown()


if __name__=="__main__":
```

```
            world_size = 3
            for num_split in [1, 2, 4, 8]:
                tik = time.time()
                mp.spawn(run_worker, args=(world_size, num_split), nprocs=world_size,
join=True)
                tok = time.time()
                print(f"number of splits = {num_split}, execution time = {tok - tik}")
```

5.11 使用异步执行实现批量 RPC 处理

本节演示如何使用 @rpc.functions.async_execution 装饰器构建批处理 RPC 应用程序，这有助于通过减少阻塞的 RPC 线程数量和合并被调用者上的 CUDA 操作来加快训练。

之前的章节已经展示了使用 torch.distributed.rpc 构建分布式训练应用程序的步骤，但它们没有详细说明在处理 rpc 请求时被调用方会发生什么。从 PyTorch v1.5 开始，每个 RPC 请求都会阻塞被调用者上的一个线程来执行该请求中的函数，直到该函数返回。这适用于许多用例，但有一点需要注意：如果用户函数在 I/O 上阻塞（例如，使用嵌套 RPC 调用），则被叫方上的 RPC 线程将不得不空闲等待，直到 I/O 完成。因此，RPC 被调用者可能会使用超出必要数量的线程。产生这个问题的原因是 RPC 将用户函数视为黑盒，并且对函数中发生的事情知之甚少。为了允许用户函数生成和释放 RPC 线程，需要向 RPC 系统提供更多提示。

自 v1.6.0 以来，PyTorch 通过引入如下两个新概念来解决这个问题。
- 一种封装异步执行的 torch.futures.Future 类型，它还支持安装回调函数。
- 一个 @rpc.functions.async_execution 装饰器，允许应用程序告诉被调用方目标函数将返回 future，并且可以在执行过程中多次暂停。

使用这两个工具，应用程序代码可以将用户函数分解为多个较小的函数，将它们作为 future 对象的回调链接在一起，并返回包含最终结果的 future。在被调用者端，当获取 future 对象时，它会安装后续的 RPC 响应准备和通信作为回调，当最终结果准备好时会触发回调。通过这种方式，被调用者不再需要阻塞一个线程并等待最终返回值。

除了减少被调用者上的空闲线程数量外，这些工具还有助于使批量 RPC 处理更容易、更快。下文演示了如何使用 @rpc.functions.async_execution 装饰器构建分布式批量更新参数服务器和批量处理强化学习应用程序。

5.11.1 批量更新参数服务器

考虑一个具有一个参数服务器和多个训练器的同步参数服务器训练应用程序。在该应用程序中,参数服务器保存参数并等待所有训练器报告梯度。在每次迭代中,它都会等待,直到接收到来自所有训练器的梯度,然后在一次拍摄中更新所有参数。在每次迭代中,它都会等待,直到接收到来自所有训练器的梯度,然后更新所有参数。下面的代码显示了参数服务器类的实现。update_and_fetch_model() 方法使用 @rpc.functions.async_execution 进行修饰,并将由训练器调用。每次调用都返回一个 future 对象,该对象将使用更新后的模型填充。大多数训练器启动的调用只是将梯度累积到 .grad 字段,然后立即返回,并在参数服务器上生成 RPC 线程。最后到达的训练器将触发优化器步骤,并消耗之前报告的所有梯度。然后,参数服务器用更新的模型设置 future_model,该模型又通过 future 对象通知其他训练器以前的所有请求,并将更新的模型发送给所有训练器。

```python
import threading
import torchvision
import torch
import torch.distributed.rpc as rpc
from torch import optim

num_classes, batch_update_size = 30, 5

class BatchUpdateParameterServer(object):
    def __init__(self, batch_update_size=batch_update_size):
        self.model = torchvision.models.resnet50(num_classes=num_classes)
        self.lock = threading.Lock()
        self.future_model = torch.futures.Future()
        self.batch_update_size = batch_update_size
        self.curr_update_size = 0
        self.optimizer = optim.SGD(self.model.parameters(), lr=0.001, momentum=0.9)
        for p in self.model.parameters():
            p.grad = torch.zeros_like(p)

    def get_model(self):
        return self.model

    @staticmethod
    @rpc.functions.async_execution
    def update_and_fetch_model(ps_rref, grads):
        # Using the RRef to retrieve the local PS instance
        self = ps_rref.local_value()
        with self.lock:
            self.curr_update_size += 1
            # accumulate gradients into .grad field
            for p, g in zip(self.model.parameters(), grads):
                p.grad += g
```

```
            # Save the current future_model and return it to make sure the
            # returned Future object holds the correct model even if another
            # thread modifies future_model before this thread returns.
            fut = self.future_model

            if self.curr_update_size >= self.batch_update_size:
                # update the model
                for p in self.model.parameters():
                    p.grad /= self.batch_update_size
                self.curr_update_size = 0
                self.optimizer.step()
                self.optimizer.zero_grad()
                # by settiing the result on the Future object, all previous
                # requests expecting this updated model will be notified and
                # the their responses will be sent accordingly.
                fut.set_result(self.model)
                self.future_model = torch.futures.Future()

        return fut
```

训练器都使用参数服务器中的同一组参数进行初始化。在每次迭代中，每个训练者首先进行向前和向后的传播，以局部生成梯度。然后，每个训练器使用 RPC 向参数服务器报告其梯度，并通过相同 RPC 请求的返回值取回更新的参数。在训练器的实现中，目标函数是否标记有 @rpc.functions.async_execution 没有区别。训练器只需使用 rpc_sync 调用 update_and_fetch_model()，rpc_sync 将阻塞训练器，直到返回更新的模型。

```
batch_size, image_w, image_h = 20, 64, 64

class Trainer(object):
    def __init__(self, ps_rref):
        self.ps_rref, self.loss_fn = ps_rref, torch.nn.MSELoss()
        self.one_hot_indices = torch.LongTensor(batch_size) \
                                    .random_(0, num_classes) \
                                    .view(batch_size, 1)

    def get_next_batch(self):
        for _ in range(6):
            inputs = torch.randn(batch_size, 3, image_w, image_h)
            labels = torch.zeros(batch_size, num_classes) \
                        .scatter_(1, self.one_hot_indices, 1)
            yield inputs.cuda(), labels.cuda()

    def train(self):
        name = rpc.get_worker_info().name
```

```
            # get initial model parameters
            m = self.ps_rref.rpc_sync().get_model().cuda()
            # start training
            for inputs, labels in self.get_next_batch():
                self.loss_fn(m(inputs), labels).backward()
                m = rpc.rpc_sync(
                    self.ps_rref.owner(),
                    BatchUpdateParameterServer.update_and_fetch_model,
                    args=(self.ps_rref, [p.grad for p in m.cpu().parameters()]),
                ).cuda()
```

可以在没有 @rpc.functions.async_execution 装饰器的情况下实现批处理。然而，这将需要在参数服务器上阻塞更多的 RPC 线程，或者使用另一轮 RPC 来获取更新的模型，后者将增加更多的代码复杂性和通信开销。

本节使用了一个简单的参数服务器训练示例来展示如何使用 @rpc.functions.async_execution 装饰器实现批处理 RPC 应用程序。下一节将重新实现 5.8 节"分布式 RPC 框架入门"中的强化学习示例。

5.11.2 批量处理 CartPole 求解器

本节以 OpenAI Gym 的 CartPole-v1 为例，展示批处理 RPC 对性能的影响。由于我们的目标是演示 @rpc.functions.async_execution 的使用，而不是构建最好的 CartPole 求解器或解决大多数不同的 RL 问题，所以，使用非常简单的策略和奖励计算策略，并专注于多观察者单代理批处理 RPC 实现。我们使用与 5.8 节"分布式 RPC 框架入门"类似的策略模型。与 5.8 节"分布式 RPC 框架入门"相比，不同之处在于它的构造函数采用了一个额外的批处理参数，该参数控制 F.softmax 的 dim 参数，因为在批处理中，正向函数中的 x 参数包含来自多个观察者的状态，所以，维度需要正确地更改，其他一切都完好无损。

```
import argparse
import torch.nn as nn
import torch.nn.functional as F

parser = argparse.ArgumentParser(description='PyTorch RPC Batch RL example')
parser.add_argument('--gamma', type=float, default=1.0, metavar='G',
                    help='discount factor (default: 1.0)')
parser.add_argument('--seed', type=int, default=543, metavar='N',
                    help='random seed (default: 543)')
parser.add_argument('--num-episode', type=int, default=10, metavar='E',
                    help='number of episodes (default: 10)')
args = parser.parse_args()
```

```python
torch.manual_seed(args.seed)

class Policy(nn.Module):
    def __init__(self, batch=True):
        super(Policy, self).__init__()
        self.affine1 = nn.Linear(4, 128)
        self.dropout = nn.Dropout(p=0.6)
        self.affine2 = nn.Linear(128, 2)
        self.dim = 2 if batch else 1

    def forward(self, x):
        x = self.affine1(x)
        x = self.dropout(x)
        x = F.relu(x)
        action_scores = self.affine2(x)
        return F.softmax(action_scores, dim=self.dim)
```

Observer 的构造函数也会相应地进行调整。Observer 还接收一个批处理参数，该参数控制 Observer 用于选择操作的代理函数。在批处理模式下，Observer 在 Agent 上调用 select_action_batch() 函数，该函数将很快出现，并且该函数将用 @rpc.functions.async_execution 进行修饰。

```python
import gym
import torch.distributed.rpc as rpc

class Observer:
    def __init__(self, batch=True):
        self.id = rpc.get_worker_info().id - 1
        self.env = gym.make('CartPole-v1')
        self.env.seed(args.seed)
        self.select_action = Agent.select_action_batch if batch else Agent.select_action
```

与 5.8 节"分布式 RPC 框架入门"相比，观察者的行为略有不同。观察者不是在环境停止时退出，而是在每一回合中运行 n_steps 次迭代。当环境返回时，观察者只需重置环境并重新开始。通过这种设计，代理将从每个观察者那里接收固定数量的状态，因此可以将它们打包到一个固定大小的张量中。在每一步中，观察者都使用 RPC 将其状态发送给代理，并通过返回值获取动作。在每回合结束时，观察者会将所有步骤的奖励返回给代理。请注意，此 run_epiode() 函数将由代理使用 RPC 调用。因此，此函数中的 rpc_sync 调用将是一个嵌套的 rpc 调用。也可以将此函数标记为 @rpc.functions.async_execution，以避免阻塞观察者上的一个线程。但是，由于瓶颈是代理而不是观察者，所以，阻塞观察者进程上的一个线程应该是可以的。

```python
import torch

class Observer:
    ...

    def run_episode(self, agent_rref, n_steps):
        state, ep_reward = self.env.reset(), NUM_STEPS
        rewards = torch.zeros(n_steps)
        start_step = 0
        for step in range(n_steps):
            state = torch.from_numpy(state).float().unsqueeze(0)
            # send the state to the agent to get an action
            action = rpc.rpc_sync(
                agent_rref.owner(),
                self.select_action,
                args=(agent_rref, self.id, state)
            )

            # apply the action to the environment, and get the reward
            state, reward, done, _ = self.env.step(action)
            rewards[step] = reward

            if done or step + 1 >= n_steps:
                curr_rewards = rewards[start_step:(step + 1)]
                R = 0
                for i in range(curr_rewards.numel() -1, -1, -1):
                    R = curr_rewards[i] + args.gamma * R
                    curr_rewards[i] = R
                state = self.env.reset()
                if start_step == 0:
                    ep_reward = min(ep_reward, step - start_step + 1)
                start_step = step + 1

        return [rewards, ep_reward]
```

Agent 的构造函数还接收一个批处理参数，该参数控制动作概率的批处理方式。在批处理模式中，saved_log_probs 包含张量列表，其中每个张量都包含一个步骤中来自所有观察者的动作概率。在没有批处理的情况下，saved_log_probs 是一个字典，其中，键是观察者 id，值是该观察者的动作概率列表。

```python
import threading
from torch.distributed.rpc import RRef

class Agent:
    def __init__(self, world_size, batch=True):
```

```python
        self.ob_rrefs = []
        self.agent_rref = RRef(self)
        self.rewards = {}
        self.policy = Policy(batch).cuda()
        self.optimizer = optim.Adam(self.policy.parameters(), lr=1e-2)
        self.running_reward = 0

        for ob_rank in range(1, world_size):
            ob_info = rpc.get_worker_info(OBSERVER_NAME.format(ob_rank))
            self.ob_rrefs.append(rpc.remote(ob_info, Observer, args=(batch,)))
            self.rewards[ob_info.id] = []

        self.states = torch.zeros(len(self.ob_rrefs), 1, 4)
        self.batch = batch
        self.saved_log_probs = [] if batch else {k:[] for k in range(len(self.ob_rrefs))}
        self.future_actions = torch.futures.Future()
        self.lock = threading.Lock()
        self.pending_states = len(self.ob_rrefs)
```

非批处理 select_acion 只需运行状态抛出策略，保存动作概率，并立即将动作返回给观察者。

```python
from torch.distributions import Categorical

class Agent:
    ...

    @staticmethod
    def select_action(agent_rref, ob_id, state):
        self = agent_rref.local_value()
        probs = self.policy(state.cuda())
        m = Categorical(probs)
        action = m.sample()
        self.saved_log_probs[ob_id].append(m.log_prob(action))
        return action.item()
```

使用批处理，状态存储在 2D 张量 self.states 中，使用观察者 id 作为行 id。然后，代理通过将回调函数安装到批量生成的 self.future_actions 对象来链接 future，该对象将填充使用该观察者的 id 索引的特定行。最后到达的观察者一次性通过策略运行所有批处理状态，并相应地设置 self.future_actions。当这种情况发生时，self.future_actions 上安装的所有回调函数都将被触发，它们的返回值将用于填充链接的 future 对象。future 对象反过来通知代理为来自其他观察者的所有以前的 RPC 请求准备和通信响应。

```python
class Agent:
    ...
```

```python
    @staticmethod
    @rpc.functions.async_execution
    def select_action_batch(agent_rref, ob_id, state):
        self = agent_rref.local_value()
        self.states[ob_id].copy_(state)
        future_action = self.future_actions.then(
            lambda future_actions: future_actions.wait()[ob_id].item()
        )

        with self.lock:
            self.pending_states -= 1
            if self.pending_states == 0:
                self.pending_states = len(self.ob_rrefs)
                probs = self.policy(self.states.cuda())
                m = Categorical(probs)
                actions = m.sample()
                self.saved_log_probs.append(m.log_prob(actions).t()[0])
                future_actions = self.future_actions
                self.future_actions = torch.futures.Future()
                future_actions.set_result(actions.cpu())
        return future_action
```

下面定义如何将不同的 RPC 函数结合在一起。代理控制着每回合的执行。首先使用 rpc_async 来启动所有观察者的回合，并阻止返回的 future，该 future 将用观察者奖励填充。然后，代理将保存的动作概率和返回的观察者奖励转换为预期的数据格式，并启动训练步骤。最后，代理重置所有状态并返回当前回合的奖励。此函数是运行一回合的入口点。

```python
class Agent:
    ...

    def run_episode(self, n_steps=0):
        futs = []
        for ob_rref in self.ob_rrefs:
            # make async RPC to kick off an episode on all observers
            futs.append(ob_rref.rpc_async().run_episode(self.agent_rref, n_steps))

        # wait until all obervers have finished this episode
        rets = torch.futures.wait_all(futs)
        rewards = torch.stack([ret[0] for ret in rets]).cuda().t()
        ep_rewards = sum([ret[1] for ret in rets]) / len(rets)

        # stack saved probs into one tensor
```

```python
        if self.batch:
            probs = torch.stack(self.saved_log_probs)
        else:
            probs = [torch.stack(self.saved_log_probs[i]) for i in range(len(rets))]
            probs = torch.stack(probs)

        policy_loss = -probs * rewards / len(rets)
        policy_loss.sum().backward()
        self.optimizer.step()
        self.optimizer.zero_grad()

        # reset variables
        self.saved_log_probs = [] if self.batch else {k:[] for k in range(len(self.ob_rrefs))}
        self.states = torch.zeros(len(self.ob_rrefs), 1, 4)

        # calculate running rewards
        self.running_reward = 0.5 * ep_rewards + 0.5 * self.running_reward
        return ep_rewards, self.running_reward
```

其余的代码是启动和日志记录的正常过程。本节中的所有观察者都被动地等待来自代理的命令。

```python
def run_worker(rank, world_size, n_episode, batch, print_log=True):
    os.environ['MASTER_ADDR'] = 'localhost'
    os.environ['MASTER_PORT'] = '29500'
    if rank == 0:
        # rank0 is the agent
        rpc.init_rpc(AGENT_NAME, rank=rank, world_size=world_size)

        agent = Agent(world_size, batch)
        for i_episode in range(n_episode):
            last_reward, running_reward = agent.run_episode(n_steps=NUM_STEPS)

            if print_log:
                print('Episode {}\tLast reward: {:.2f}\tAverage reward: {:.2f}'.format(
                    i_episode, last_reward, running_reward))
    else:
        # other ranks are the observer
        rpc.init_rpc(OBSERVER_NAME.format(rank), rank=rank, world_size=world_size)
        # observers passively waiting for instructions from agents
    rpc.shutdown()
```

```
def main():
    for world_size in range(2, 12):
        delays = []
        for batch in [True, False]:
            tik = time.time()
            mp.spawn(
                run_worker,
                args=(world_size, args.num_episode, batch),
                nprocs=world_size,
                join=True
            )
            tok = time.time()
            delays.append(tok - tik)

        print(f"{world_size}, {delays[0]}, {delays[1]}")

if __name__ == '__main__':
    main()
```

批量 RPC 有助于将操作动作推理为较少的 CUDA 操作，从而减少了摊销的开销。

5.12 分布式数据并行与分布式 RPC 框架的结合

本节使用一个简单的示例来演示如何将分布式数据并行（Distributed Data Parallel，DDP）与分布式 RPC 框架相结合，以将分布式数据并行性与分布式模型并行性相结合，从而训练一个简单模型。

"5.4 分布式数据并行入门"和"5.8 分布式 RPC 框架入门"分别描述了如何执行分布式数据并行和分布式模型并行训练。尽管如此，有几种训练模式可能需要将这两种技术结合起来。例如：

（1）如果有一个具有稀疏部分（大嵌入表）和密集部分（FC 层）的模型，可能希望将嵌入表放在参数服务器上，并使用 DistributedDataParallel 在多个训练器之间复制 FC 层。分布式 RPC 框架可用于在参数服务器上执行嵌入查找。

（2）可以使用分布式 RPC 框架在多个工作程序之间流水线化模型的各个阶段，并使用 DistributedDataParallel 复制每个阶段（如果需要）。

本节将介绍上述情况（1）。总共有 4 个工作者，如下所示：

（1）一个主服务器，负责在参数服务器上创建嵌入表（nn.EmbeddingBag）。主服务器还会在两个训练器上驱动训练循环。

（2）一个参数服务器，它将嵌入表保存在内存中，并响应来自主服务器和训练器的 RPC。

（3）两个训练器，用于存储 FC 层（nn.Lineral），该层使用 DistributedDataParallel 在它们之间进行复制。训练器还负责执行前向传播、后向传播和优化步骤。

整个训练过程执行如下：

（1）主服务器创建一个 RemoteModule，该 RemoteModule 在参数服务器上保存一个嵌入表。

（2）主服务器启动训练器的训练循环，并将远程模块传递给训练器。

（3）训练器创建一个 HybridModel，该模型首先使用主服务器提供的远程模块执行嵌入查找，然后执行封装在 DDP 中的 FC 层。

（4）训练器执行模型的正向传播，并利用损失使用分布式 Autograd 执行反向传递。

（5）作为向后遍历的一部分，将首先计算 FC 层的梯度，并通过 DDP 中的 allreduce 将其同步到所有训练器。

（6）分布式 Autograd 将梯度传播到参数服务器，在该服务器中更新嵌入表的梯度。

（7）分布式优化器用于更新所有参数。

现在，来详细了解每一部分。首先，在进行任何训练之前，需要设置所有工作者。这里创建了 4 个进程，使等级 0 和 1 是训练器，等级 2 是主进程，等级 3 是参数服务器。

使用 TCP init_method 在 4 个工作者上初始化 RPC 框架。RPC 初始化完成后，主服务器将使用 RemoteModule 创建一个远程模块，该模块在参数服务器上保存 EmbeddingBag 层。然后，主服务器通过使用 rpc_async 在每个训练器上调用 _run_trainer，循环遍历每个训练器并开始训练循环。最后，主服务器等待所有训练结束后退出。

训练器首先使用 init_process_group 为 world_size=2 的 DDP 初始化一个 ProcessGroup。接下来，它们使用 TCP init_method 初始化 RPC 框架。注意，RPC 初始化和 ProcessGroup 初始化中的端口不同。这是为了避免两个框架初始化之间的端口冲突。初始化完成后，训练器只需等待来自主服务器的 _run_trainer RPC。

参数服务器只是初始化 RPC 框架，并等待来自训练器和主服务器的 RPC。

```
def run_worker(rank, world_size):
    r"""
    A wrapper function that initializes RPC, calls the function, and shuts down
    RPC.
    """

    # We need to use different port numbers in TCP init_method for init_rpc and
    # init_process_group to avoid port conflicts.
    rpc_backend_options = TensorPipeRpcBackendOptions()
    rpc_backend_options.init_method = "tcp://localhost:29501"
```

```python
    # Rank 2 is master, 3 is ps and 0 and 1 are trainers.
    if rank == 2:
        rpc.init_rpc(
            "master",
            rank=rank,
            world_size=world_size,
            rpc_backend_options=rpc_backend_options,
        )

        remote_emb_module = RemoteModule(
            "ps",
            torch.nn.EmbeddingBag,
            args=(NUM_EMBEDDINGS, EMBEDDING_DIM),
            kwargs={"mode": "sum"},
        )

        # Run the training loop on trainers.
        futs = []
        for trainer_rank in [0, 1]:
            trainer_name = "trainer{}".format(trainer_rank)
            fut = rpc.rpc_async(
                trainer_name, _run_trainer, args=(remote_emb_module, trainer_rank)
            )
            futs.append(fut)

        # Wait for all training to finish.
        for fut in futs:
            fut.wait()
    elif rank <= 1:
        # Initialize process group for Distributed DataParallel on trainers.
        dist.init_process_group(
                backend="gloo", rank=rank, world_size=2, init_method="tcp://localhost:29500"
        )

        # Initialize RPC.
        trainer_name = "trainer{}".format(rank)
        rpc.init_rpc(
            trainer_name,
            rank=rank,
            world_size=world_size,
            rpc_backend_options=rpc_backend_options,
        )

        # Trainer just waits for RPCs from master.
```

```
        else:
            rpc.init_rpc(
                "ps",
                rank=rank,
                world_size=world_size,
                rpc_backend_options=rpc_backend_options,
            )
            # parameter server do nothing
            pass

        # block until all rpcs finish
        rpc.shutdown()


if __name__ == "__main__":
    # 2 trainers, 1 parameter server, 1 master.
    world_size = 4
    mp.spawn(run_worker, args=(world_size,), nprocs=world_size, join=True)
```

在讨论训练器的详细信息之前，先介绍一下训练器使用的 HybridModel。使用在参数服务器和用于 DDP 的设备上保存嵌入表的远程模块来初始化 HybridModel。模型的初始化在 DDP 中包装了 nn.Linear 层，以在所有训练器之间复制和同步该层。

该模型的正向方法非常简单。它使用 RemoteModule 的正向在参数服务器上执行嵌入查找，并将其输出传递到 FC 层。

```
class HybridModel(torch.nn.Module):
    r"""
    The model consists of a sparse part and a dense part.
        1) The dense part is an nn.Linear module that is replicated across all trainers using DistributedDataParallel.
        2) The sparse part is a Remote Module that holds an nn.EmbeddingBag on the parameter server.
        This remote model can get a Remote Reference to the embedding table on the parameter server.
    """

    def __init__(self, remote_emb_module, device):
        super(HybridModel, self).__init__()
        self.remote_emb_module = remote_emb_module
        self.fc = DDP(torch.nn.Linear(16, 8).cuda(device), device_ids=[device])
        self.device = device

    def forward(self, indices, offsets):
        emb_lookup = self.remote_emb_module.forward(indices, offsets)
```

```
        return self.fc(emb_lookup.cuda(self.device))
```

接下来，介绍训练器上的设置。训练器首先使用远程模块创建上述 HybridModel，该远程模块保存参数服务器上的嵌入表及其自己的等级。

现在，需要检索要使用 DistributedOptimizer 优化的所有参数的 RRef 列表。要从参数服务器中检索嵌入表的参数，可以调用 RemoteModule 的 remote_parameters，它可以遍历嵌入表的所有参数，并返回一个 RRef 列表。训练器通过 RPC 在参数服务器上调用此方法，以接收所需参数的 RRef 列表。由于 DistributedOptimizer 总是将 RRef 列表带到需要优化的参数，所以，需要为 FC 层的本地参数创建 RRef。这是通过遍历 model.fc.parameters() 来完成的，为每个参数创建一个 RRef，并将其附加到从 remote_parameters() 返回的列表中。注意，不能使用 model.parameters()，因为它将递归调用 RemoteModule 不支持的 model.remote_emb_module.parameters()。

最后，使用所有 RRef 创建 DistributedOptimizer，并定义 CrossEntropyLoss 函数。

```
def _run_trainer(remote_emb_module, rank):
    r"""
    Each trainer runs a forward pass which involves an embedding lookup on the
    parameter server and running nn.Linear locally. During the backward pass,
    DDP is responsible for aggregating the gradients for the dense part
    (nn.Linear) and distributed autograd ensures gradients updates are
    propagated to the parameter server.
    """

    # Setup the model.
    model = HybridModel(remote_emb_module, rank)

    # Retrieve all model parameters as rrefs for DistributedOptimizer.

    # Retrieve parameters for embedding table.
    model_parameter_rrefs = model.remote_emb_module.remote_parameters()

    # model.fc.parameters() only includes local parameters.
    # NOTE: Cannot call model.parameters() here,
    # because this will call remote_emb_module.parameters(),
    # which supports remote_parameters() but not parameters().
    for param in model.fc.parameters():
        model_parameter_rrefs.append(RRef(param))

    # Setup distributed optimizer
    opt = DistributedOptimizer(
        optim.SGD,
        model_parameter_rrefs,
        lr=0.05,
```

```
)
criterion = torch.nn.CrossEntropyLoss()
```

下面介绍在每个训练器上运行的主要训练循环。get_next_batch() 是一个辅助函数,用于生成用于训练的随机输入和目标。为多个回合和每个批次运行训练循环:

(1) 为分布式 Autograd 设置分布式 Autograd 上下文。
(2) 运行模型的正向传播并检索其输出。
(3) 使用损失函数,根据输出和目标计算损失。
(4) 使用分布式 Autograd 执行分布式反向传播。
(5) 运行分布式优化器步骤来优化所有参数。

```
def get_next_batch(rank):
    for _ in range(10):
        num_indices = random.randint(20, 50)
        indices = torch.LongTensor(num_indices).random_(0, NUM_EMBEDDINGS)

        # Generate offsets.
        offsets = []
        start = 0
        batch_size = 0
        while start < num_indices:
            offsets.append(start)
            start += random.randint(1, 10)
            batch_size += 1

        offsets_tensor = torch.LongTensor(offsets)
        target = torch.LongTensor(batch_size).random_(8).cuda(rank)
        yield indices, offsets_tensor, target

# Train for 100 epochs
for epoch in range(100):
    # create distributed autograd context
    for indices, offsets, target in get_next_batch(rank):
        with dist_autograd.context() as context_id:
            output = model(indices, offsets)
            loss = criterion(output, target)

            # Run distributed backward pass
            dist_autograd.backward(context_id, [loss])

            # Tun distributed optimizer
            opt.step(context_id)
```

```
                    # Not necessary to zero grads as each iteration creates a different
                    # distributed autograd context which hosts different grads
        print("Training done for epoch {}".format(epoch))
```

5.13 使用流水线并行性训练 transformer 模型

本节将把一个 transformer 模型拆分到两个 GPU 上,并使用流水线并行性来训练模型。最大数量的参数属于 nn.TransformerEncoder 层。nn.TransformerEncoder 由多层 nn.TransformerEncoderLayer 组成。因此,重点介绍 nn.TransformerEncoder。我们对模型进行了拆分,使得一半 nn.TransformerEncoderLayer 位于一个 GPU 上,另一半位于另一个 GPU 上。要做到这一点,需要将编码器和解码器部分提取到单独的模块中,然后构建 nn.Sequential 表示原始 transformer 模块。

```python
import sys
import math
import torch
import torch.nn as nn
import torch.nn.functional as F
import tempfile
from torch.nn import TransformerEncoder, TransformerEncoderLayer

if sys.platform == 'win32':
    print('Windows platform is not supported for pipeline parallelism')
    sys.exit(0)
if torch.cuda.device_count() < 2:
    print('Need at least two GPU devices for this tutorial')
    sys.exit(0)

class Encoder(nn.Module):
    def __init__(self, ntoken, ninp, dropout=0.5):
        super(Encoder, self).__init__()
        self.pos_encoder = PositionalEncoding(ninp, dropout)
        self.encoder = nn.Embedding(ntoken, ninp)
        self.ninp = ninp
        self.init_weights()

    def init_weights(self):
        initrange = 0.1
        self.encoder.weight.data.uniform_(-initrange, initrange)

    def forward(self, src):
```

```
            # Need (S, N) format for encoder.
            src = src.t()
            src = self.encoder(src) * math.sqrt(self.ninp)
            return self.pos_encoder(src)

    class Decoder(nn.Module):
        def __init__(self, ntoken, ninp):
            super(Decoder, self).__init__()
            self.decoder = nn.Linear(ninp, ntoken)
            self.init_weights()

        def init_weights(self):
            initrange = 0.1
            self.decoder.bias.data.zero_()
            self.decoder.weight.data.uniform_(-initrange, initrange)

        def forward(self, inp):
            # Need batch dimension first for output of pipeline.
            return self.decoder(inp).permute(1, 0, 2)
```

PositionalEncoding 模块注入一些关于序列中表征的相对或绝对位置的信息。位置编码与嵌入具有相同的维度，因此，可以将两者相加。这里，使用不同频率的正弦和余弦函数。

```
    class PositionalEncoding(nn.Module):

        def __init__(self, d_model, dropout=0.1, max_len=5000):
            super(PositionalEncoding, self).__init__()
            self.dropout = nn.Dropout(p=dropout)

            pe = torch.zeros(max_len, d_model)
            position = torch.arange(0, max_len, dtype=torch.float).unsqueeze(1)
            div_term = torch.exp(torch.arange(0, d_model, 2).float() * (-math.log(10000.0)
/ d_model))
            pe[:, 0::2] = torch.sin(position * div_term)
            pe[:, 1::2] = torch.cos(position * div_term)
            pe = pe.unsqueeze(0).transpose(0, 1)
            self.register_buffer('pe', pe)

        def forward(self, x):
            x = x + self.pe[:x.size(0), :]
            return self.dropout(x)
```

训练过程使用来自 torchtext 的 Wikitext-2 数据集。若要访问 torchtext 数据集，需安装 torchdata。vocab 对象是基于 train 数据集构建的，用于将表征数字化为张量。从顺序数据开始，batchify() 函数将数据集排列成列，将数据划分为 batch_size 大小的批后，修剪掉所有剩余的表征。

```python
import torch
from torchtext.datasets import WikiText2
from torchtext.data.utils import get_tokenizer
from torchtext.vocab import build_vocab_from_iterator

train_iter = WikiText2(split='train')
tokenizer = get_tokenizer('basic_english')
vocab = build_vocab_from_iterator(map(tokenizer, train_iter), specials=["<unk>"])
vocab.set_default_index(vocab["<unk>"])

def data_process(raw_text_iter):
    data = [torch.tensor(vocab(tokenizer(item)), dtype=torch.long) for item in raw_text_iter]
    return torch.cat(tuple(filter(lambda t: t.numel() > 0, data)))

train_iter, val_iter, test_iter = WikiText2()
train_data = data_process(train_iter)
val_data = data_process(val_iter)
test_data = data_process(test_iter)

device = torch.device("cuda")

def batchify(data, bsz):
    # Divide the dataset into ``bsz`` parts.
    nbatch = data.size(0) // bsz
    # Trim off any extra elements that wouldn't cleanly fit (remainders).
    data = data.narrow(0, 0, nbatch * bsz)
    # Evenly divide the data across the ``bsz`` batches.
    data = data.view(bsz, -1).t().contiguous()
    return data.to(device)

batch_size = 20
eval_batch_size = 10
train_data = batchify(train_data, batch_size)
val_data = batchify(val_data, eval_batch_size)
test_data = batchify(test_data, eval_batch_size)
```

get_batch() 函数为 transformer 模型生成输入和目标序列。它将源数据细分为长度为 bptt 的块。

```python
bptt = 25
def get_batch(source, i):
```

```
        seq_len = min(bptt, len(source) - 1 - i)
        data = source[i:i+seq_len]
        target = source[i+1:i+1+seq_len].view(-1)
        # Need batch dimension first for pipeline parallelism.
        return data.t(), target
```

为了演示使用流水线并行性训练大型 transformer 模型,适当放大了 transformer 层。我们使用 4096 的嵌入维度,4096 的隐藏大小,16 个注意力头和 12 个 transformer 层。这将创建一个具有约 14 亿个参数的模型。

需要初始化 RPC 框架,因为管道通过 RRef 依赖于 RPC 框架,这允许将来扩展到跨主机管道。由于使用单个进程来驱动多个 GPU,所以,需要仅使用单个工作者来初始化 RPC 框架。

然后用一个 GPU 上的 8 个 transformer 层和另一个 GPU 上的 8 个 transformer 层来初始化流水线。

```
    ntokens = len(vocab) # the size of vocabulary
    emsize = 4096 # embedding dimension
    nhid = 4096 # the dimension of the feedforward network model in ``nn.TransformerEncoder``
    nlayers = 12 # the number of ``nn.TransformerEncoderLayer`` in ``nn.TransformerEncoder``
    nhead = 16 # the number of heads in the Multihead Attention models
    dropout = 0.2 # the dropout value

    from torch.distributed import rpc
    tmpfile = tempfile.NamedTemporaryFile()
    rpc.init_rpc(
        name="worker",
        rank=0,
        world_size=1,
        rpc_backend_options=rpc.TensorPipeRpcBackendOptions(
            init_method="file://{}".format(tmpfile.name),
            # Specifying _transports and _channels is a workaround and we no longer
            # will have to specify _transports and _channels for PyTorch
            # versions >= 1.8.1
            _transports=["ibv", "uv"],
            _channels=["cuda_ipc", "cuda_basic"],
        )
    )

    num_gpus = 2
    partition_len = ((nlayers - 1) // num_gpus) + 1
```

```python
# Add encoder in the beginning.
tmp_list = [Encoder(ntokens, emsize, dropout).cuda(0)]
module_list = []

# Add all the necessary transformer blocks.
for i in range(nlayers):
    transformer_block = TransformerEncoderLayer(emsize, nhead, nhid, dropout)
    if i != 0 and i % (partition_len) == 0:
        module_list.append(nn.Sequential(*tmp_list))
        tmp_list = []
    device = i // (partition_len)
    tmp_list.append(transformer_block.to(device))

# Add decoder in the end.
tmp_list.append(Decoder(ntokens, emsize).cuda(num_gpus - 1))
module_list.append(nn.Sequential(*tmp_list))

from torch.distributed.pipeline.sync import Pipe

# Build the pipeline.
chunks = 8
model = Pipe(torch.nn.Sequential(*module_list), chunks = chunks)


def get_total_params(module: torch.nn.Module):
    total_params = 0
    for param in module.parameters():
        total_params += param.numel()
    return total_params

print ('Total parameters in model: {:,}'.format(get_total_params(model)))
```

交叉熵损失用于跟踪损失，SGD 实现随机梯度下降法作为优化器。将初始学习率设置为 5.0。StepLR 用于通过纪元来调整学习率。在训练过程中，使用 nn.utils.clip_grad_norm_() 函数将所有梯度缩放到一起，以防止梯度爆炸。

```python
criterion = nn.CrossEntropyLoss()
lr = 5.0 # learning rate
optimizer = torch.optim.SGD(model.parameters(), lr=lr)
scheduler = torch.optim.lr_scheduler.StepLR(optimizer, 1.0, gamma=0.95)

import time
def train():
    model.train() # Turn on the train mode
    total_loss = 0.
```

```python
    start_time = time.time()
    ntokens = len(vocab)

    # Train only for 50 batches to keep script execution time low.
    nbatches = min(50 * bptt, train_data.size(0) - 1)

    for batch, i in enumerate(range(0, nbatches, bptt)):
        data, targets = get_batch(train_data, i)
        optimizer.zero_grad()
        # Since the Pipe is only within a single host and process the ``RRef``
        # returned by forward method is local to this node and can simply
        # retrieved via ``RRef.local_value()``.
        output = model(data).local_value()
        # Need to move targets to the device where the output of the
        # pipeline resides.
        loss = criterion(output.view(-1, ntokens), targets.cuda(1))
        loss.backward()
        torch.nn.utils.clip_grad_norm_(model.parameters(), 0.5)
        optimizer.step()

        total_loss += loss.item()
        log_interval = 10
        if batch % log_interval == 0 and batch > 0:
            cur_loss = total_loss / log_interval
            elapsed = time.time() - start_time
            print('| epoch {:3d} | {:5d}/{:5d} batches | '
                  'lr {:02.2f} | ms/batch {:5.2f} | '
                  'loss {:5.2f} | ppl {:8.2f}'.format(
                    epoch, batch, nbatches // bptt, scheduler.get_lr()[0],
                    elapsed * 1000 / log_interval,
                    cur_loss, math.exp(cur_loss)))
            total_loss = 0
            start_time = time.time()

def evaluate(eval_model, data_source):
    eval_model.eval() # Turn on the evaluation mode
    total_loss = 0.
    ntokens = len(vocab)
    # Evaluate only for 50 batches to keep script execution time low.
    nbatches = min(50 * bptt, data_source.size(0) - 1)
    with torch.no_grad():
        for i in range(0, nbatches, bptt):
            data, targets = get_batch(data_source, i)
            output = eval_model(data).local_value()
            output_flat = output.view(-1, ntokens)
```

```
                # Need to move targets to the device where the output of the
                # pipeline resides.
                total_loss += len(data) * criterion(output_flat, targets.cuda(1)).item()
    return total_loss / (len(data_source) - 1)
```

在回合上循环。如果验证损失是迄今为止看到的最好的,则保存模型。在每个回合之后调整学习率。

```
best_val_loss = float("inf")
epochs = 3 # The number of epochs
best_model = None

for epoch in range(1, epochs + 1):
    epoch_start_time = time.time()
    train()
    val_loss = evaluate(model, val_data)
    print('-' * 89)
    print('| end of epoch {:3d} | time: {:5.2f}s | valid loss {:5.2f} | '
          'valid ppl {:8.2f}'.format(epoch, (time.time() - epoch_start_time),
                                      val_loss, math.exp(val_loss)))
    print('-' * 89)

    if val_loss < best_val_loss:
        best_val_loss = val_loss
        best_model = model

    scheduler.step()
```

应用最佳模型来检查测试数据集的结果。

```
test_loss = evaluate(best_model, test_data)
print('=' * 89)
print('| End of training | test loss {:5.2f} | test ppl {:8.2f}'.format(
    test_loss, math.exp(test_loss)))
print('=' * 89)
```

5.14 本章小结

本章介绍了使用 PyTorch 实现分布式训练的方法。特别是介绍了分布式数据并行入门,以及使用流水线并行性训练 transformer 模型。